The Shadow Owners' Companion

Maintenance Projects for Rolls-Royce Silver Shadow and Bentley T Enthusiasts

Jon J. Waples

SHERBOURNE MEWS, LLC

The Shadow Owners' Companion
by Jon J. Waples
First Edition, v5

ISBN 978-0-6151-4103-9

Sherbourne Mews, LLC
3231 Sherbourne Road
Detroit, MI 48221-1814
USA

website: www.sherbournemews.com
e-mail: sales@sherbournemews.com
phone: +1-248-327-4330

Accuracy is important to us at Sherbourne Mews. Please consult www.sherbournemews.com/downloads to view the up to the minute errata for this document.

Contents

Preface

<u>*Shadow People*</u>

For people like us, the satisfaction of owning a Rolls-Royce Silver Shadow, Bentley T, or one of the coachbuilt derivatives is obvious yet hard to explain.

We enjoy working with our hands and troubleshooting, and the Shadow is a clear challenge. For Rolls-Royce, the Shadow was the most ambitious attempt to best the competition at any time since WWII. It is dense with period technology: hydraulic height control, an early type of anti-skid control, monocoque construction, independent suspension, an all aluminum engine, and even automatic air conditioning later in the model run.

Because all of these mechanisms compete with one another for underhood space, seemingly simple acts such as changing the spark plug wires or replacing the thermostat become "rather involved." A perfectly running Shadow is a towering achievement for the individual who maintains her, yet only those "in the know" really understand.

There is an intellectual challenge as well. The design philosophy of this car differs quite a bit from the mainstream. Troubleshooting requires an understanding of the underlying principles rather than a reliance on conventional tips and tricks.

Of course, there is more to the ownership experience than maintenance. When driving a well maintained Silver Shadow, everything is at once harmonious and hushed, significant and discreet. As you take it all in, your sense of urgency washes away: very satisfying indeed, all the way around.

<u>*Why You Need This Book*</u>

So you believe in the car. What about this book? There are three other service related documents for the Silver Shadow and Bentley T: the Handbook (Owner's Manual), the Workshop Manual, and *From the Shadow's Corner* by Cal West. I have a copy of each and so should you. What follows in these pages are all of the maintenance projects that I still had to figure out for myself during the first five years of my ownership.

This book was conceived as a companion to the previously mentioned works. The maintenance procedures that follow fill the gaps so that an enthusiastic owner can maintain his or her Shadow with only occasional help from a professional mechanic. For the most part, the projects in this book were included for one of two reasons: there is a simpler way to perform the desired task with modern tools, or there is a need to fill in detail that seems to be missing, but was obvious to a seasoned mechanic in the 1970s.

If you own a 1966-1980 Rolls-Royce Silver Shadow, Silver Shadow LWB, Silver Shadow II, Silver Wraith II, Two Door Saloon, Drophead Coupé, Corniche, Camargue, Bentley T or T2 you can buy this book for far less than one full tank of gasoline. What are you waiting for?

Replacement Parts

Occasionally, this book suggests replacement parts that are not Crewe Genuine Parts™. Most often, these suggestions are identical to the Crewe specified version, coming directly from the original manufacturer at a lower price.

Maintaining the integrity of the original design with proper service and parts honors the hard working people that created and built your car. Anything less is an insult.

Disclaimer

You should be able to safely perform the procedures described herein provided that you have reasonable mechanical ability, the right tools for the job, and a fair dose of common sense. I have personally performed all of the maintenance procedures described in this book. Even with the right tools and common sense at my disposal, I have managed on occasion to cut my hands, burn my fingers, ruin paint, and utter an expletive or two, usually when I was rushing or frustrated. Take your time and think about what you are doing!

Please forgive me for reminding you that your motor car deserves respect as you work on it: it can crush, burn, scald, cut, bruise, poison, asphyxiate, maim or quite possibly kill you and those nearby if you do not follow proper safety precautions. In addition, an improperly maintained motor car can catch fire or fail to start, go, turn, or stop leading to all manner of death and destruction to people and property.

The chemicals you use in and on your car, such as antifreeze, oil, gasoline, brake fluid, and brake cleaner are toxic to humans and pets and bad for the environment. Use, store and dispose of them with care.

Finally, this book is a *companion* to your Rolls-Royce Handbook and Rolls-Royce Workshop Manual and is *not* a comprehensive substitute. As such, successful completion of all the procedures in this book is *not* a guarantee that your ownership experience will be trouble free.

Nomenclature

All references to position on this vehicle are made as if you were seated inside. For example, the brake reservoir is on the left side in front of the engine bulkhead. The Rolls-Royce Workshop Manual gets into nearside and offside relative to the driver's position but the nearside happens to be on the left hand (curb) side in England–too confusing for the television-addled American mind. When referring to the V-8 engine specifically, the "A" bank is on the right side of the car and the "B" bank on the left. Further, each cylinder is numbered front to back A1-A4 and B1-B4. Other terms that require translation:

BRITISH	AMERICAN
Arch	Wheel Arch
Backlight	Rear window
Bonnet	Hood
Boot	Trunk
Bulkhead	Firewall
Earth	Ground
Fascia	Dash
Hood	Convertible Top
Jubilee® Clip	Hose Clamp*
Loom	Wiring Harness
Methylated Spirits	Denatured Alcohol
Paraffin	Kerosene
Pressing	Sheet Metal Stamping
Saloon	Sedan
Sill	Rocker Panel
Spanner	Wrench
Tommy Bar	Tire Iron
Valance	Fascia**
Wing	Fender

---

* Jubilee clips have a non-perforated band as compared to ordinary hose clamps.

** This is the panel behind and below the front or rear bumper.

Acknowledgements

Many thanks to Greg Albers, Mark Albers, Rick Barrett, Paul Bearman, Al Brightman, Ken Burgess, Esq., Garry Campbell, George and Carlen Colgett, Dan Docherty, Andrew Dunn, Paul Fisher, Dave Hollister, Jeff Klas, Jack Miller, Scott Mueller, Todd Nagler, Bram Over, Peter Pesce, Rusty Price, Gary Rock, Bill Rohrer, Dave Sjolund, John Tupper, Richard Vaughan, Chris Whicker and Dr. Doug Wolford for their literary and technical help.

Jon J. Waples
Sherbourne Mews, Detroit
December 18, 2006

Chapter 1
Getting Started

Section 1-1
Model Nomenclature and Identification

<u>*Model Nomenclature*</u>

The Rolls-Royce Silver Shadow and Bentley T were introduced at the Earls Court Motor Show in the fall of 1965 with deliveries commencing in 1966. Coachbuilt two door coupes by James Young and Rolls-Royce owned Mulliner Park Ward became available soon thereafter. Long wheelbase versions of the Rolls-Royce Silver Shadow and Bentley T, which consisted of a four inch stretch to the rear door opening, were introduced in 1969.

The two door models were lightly revised in 1971 and renamed Corniche. In 1975, Rolls-Royce introduced a Pininfarina designed two door luxury flagship called Camargue. A major model line refresh occurred in 1977 with the Rolls-Royce Silver Shadow II, Silver Wraith II and Bentley T2.

All of these models were built on the same chassis. Technical improvements almost always found their way into the two door variants six to twelve months ahead of a broad introduction on the four door saloons. The four door models were replaced in 1981 by the Rolls-Royce Silver Spirit, Silver Spur and Bentley Mulsanne. The Camargue and Corniche lived on with revised underpinnings largely identical to the Spirit and Spur.*

Figure 1-1-1 – 1967 Rolls-Royce Silver Shadow Mulliner Park Ward Two Door Saloon, CRX2584

Figure 1-1-2 – 1967 Bentley T, SBH3097

Figure 1-1-3 – 1979 Rolls-Royce Silver Wraith II, LRK37829

---

* Because the post-1980 Camargue, Corniche and later Continental are essentially Silver Spirits underneath, they are not within the scope of this book.

Chassis Numbering

No book on Rolls-Royce and Bentley is complete without a primer on chassis numbering. When corresponding with your Crewe parts purveyor, it is important to quote your chassis number rather than model year. Since Rolls-Royce made engineering changes throughout each model year, this will ensure that you receive the correct parts.

The chassis number consists of three letters and a unique four or five digit serial number. The first two letters describe the body type and marque, respectively. The third letter was originally used to designate "home" versus export vehicles which determined which side of the car the driver controls were placed. Starting in 1972, the third letter was used to denote the model year on export models destined for North America. (See fig. 1-1-4.)

S B H 1 0 0 1

Body Type	Marque	Steering Position/Year	Sequence Number
S Saloon	R Rolls-Royce	H Home (right hand drive)	Shadow I and Derivatives:
C Coachbuilt (including Drophead Coupé prior to 6646)	B Bentley	X Export (left hand drive)	1001-4548*
D Drophead Coupé from 6646		North American Specification from 1972 (instead of X):	5001-5603
L Long Wheelbase			6001-8861
J Camargue			9001-26708
P Phantom VI		A 1972	Shadow II and Derivatives:
		B 1973	30001-41686
		C 1974	Corniche and Camargue with Mineral Oil Hydraulics:
		D 1975	50001-50776
		E 1976	
		F 1977	
		G 1978	
		K 1979	
		L 1980	

*Except 1500, 4108, 4180-4182, 4295-4297, 4503, 4504, and 4536

Figure 1-1-4 – Chassis numbering scheme.

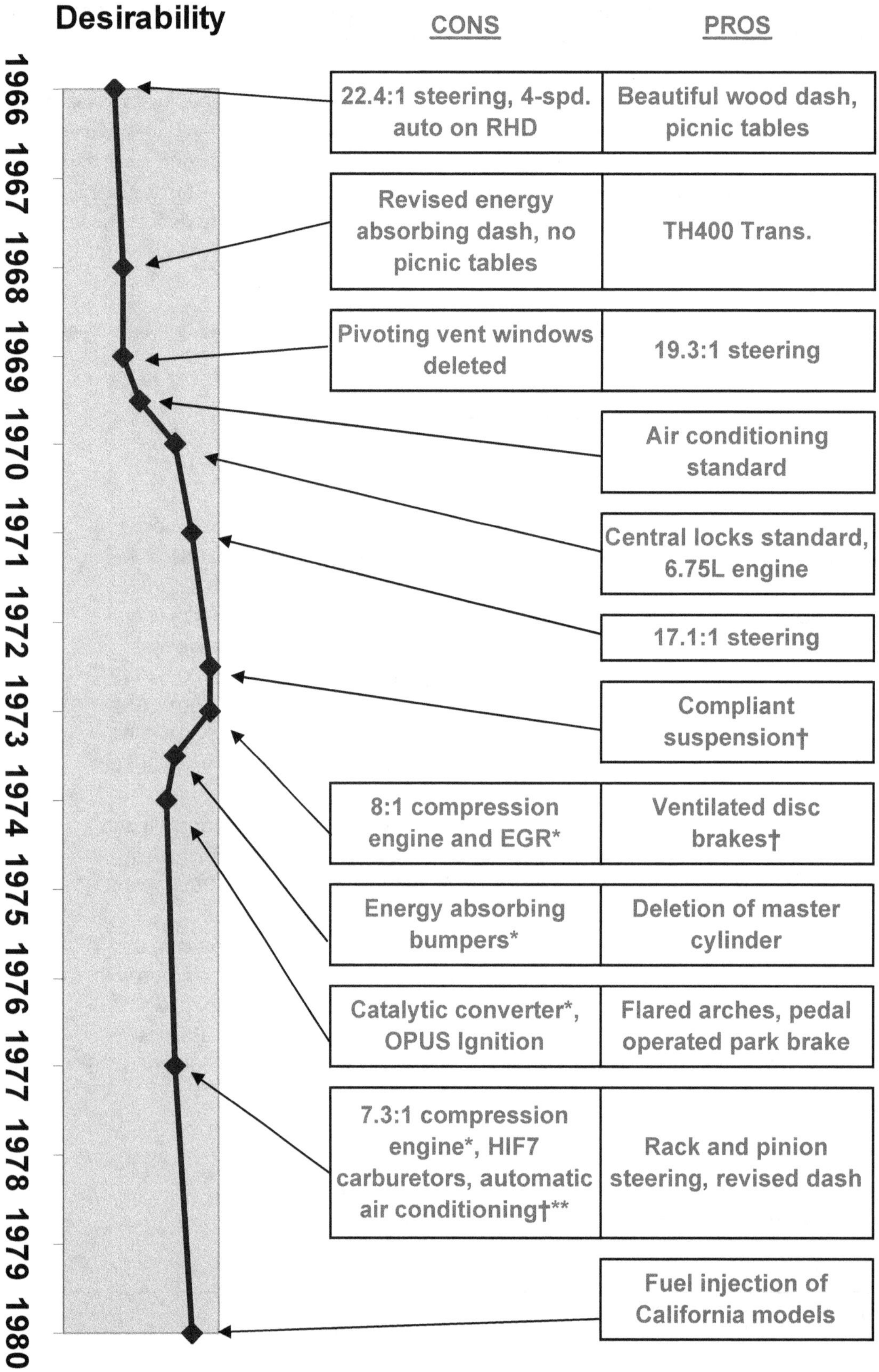

Figure 1-1-7 – Relative desirability by model year.

†Change dates are for four door saloon model - Corniche and Camargue typically sooner,
**U.S. Models, **Works great, hard to service*

The Best Years

Relative to most marques, especially those produced during the early emissions years, there are no "bad" years for the Rolls-Royce Silver Shadow and related models. Power losses due to ever tightening emissions regulations were partially, if not completely, offset with induction and exhaust system improvements.

If you are partial to great expanses of wood veneer, then the very early Shadows are best. If you prefer a "set and forget" air conditioning system and improved handling, then the Shadow II is for you.

Fig. 1-1-7 charts the major changes over the 15 year model run, with a rough guide to relative desirability.

Figure 1-1-5 – 1973 Rolls-Royce Corniche, DRB15802

Figure 1-1-6 – 1982 Rolls-Royce Camargue, JAC-05266

Section 1-2
Buying Criteria

The largest influence that a prospective owner can have on his or her future maintenance cost is purchasing the right car in the first place. Take a methodical approach. Once you take your new baby home, your fate is sealed. When it comes to any collector car purchase, buy the nicest example you can afford and get as much maintenance history as you can. I have heard this often and it is very true for the Silver Shadow series. The following is a high level summary of things to check when evaluating a Silver Shadow series car.

Body

- Identify areas of hidden rust and accident damage, especially the bottom six inches of the body.
- Look for jacking damage to the rear subframe, check the alignment of the driveshaft to the body and subframe.
- Identify any missing trim pieces.

Suspension and Steering

- Identify hidden damage to the subframes, rusty rear spring perches and worn shock absorbers and mounts.
- Verify the integrity of the suspension over rough roads. (Listen for clunking and creaking.)
- Check for excessive play in the steering wheel.
- Check for unusual wear on the tires.

Air conditioning

- Verify cooling capability both at a stop and at cruising speed with engine warm.
- Verify that all air handling doors (a.k.a. "flaps") open and close. (See Section 8-4.)

Interior

- Identify any missing trim pieces.
- Assess the condition of the floor pan, especially the front foot wells.

Powertrain

- Find evidence of proper preventive maintenance, particularly with regard to cooling and lubrication.
- Verify that cold starting, driveability, noise, vibration, shift quality and power are acceptable. A properly maintained Silver Shadow will perform like a modern luxury SUV with regard to these attributes.
- Verify the stability of the cooling system.
- Verify a hot restart after letting the car sit for 20 minutes after your test drive. Shadows (especially later ones) are not the greatest at hot starts. Just make sure that it starts.

Hydraulics

- Look for evidence of proper preventive maintenance, such as clean brake fluid in the reservoir.

- Verify functioning rear height control. (Sit in the car while it is parked with the engine running. The height control should compensate instantly by raising the rear of the car to the proper ride height. See Section 1-3, *Suspension and Brakes*.)
- Check hydraulic system function. (See Section 2-8.)

When evaluating a used car, you are evaluating both the car and its owner. The following photographs illustrate areas of particular note. Do not be timid while inspecting. Wear expendable clothes, use a flashlight and crawl underneath!

Bumpers tend to rust from the inside out, particularly around the mounting hardware.

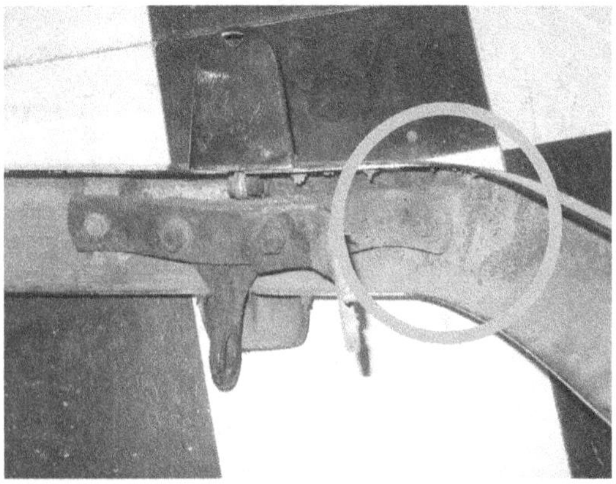

Rear bumpers rust at the corners because they tend to catch flying debris from the road wheels.

The door sills and wheel arches are common locations to find rot.

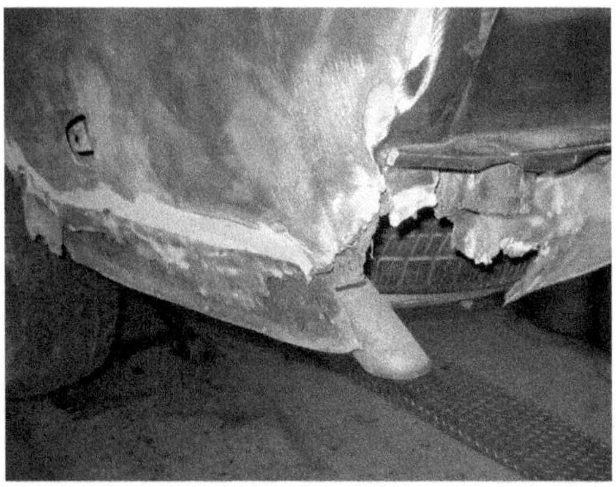

In addition, the rear valance is also rot prone as it tends to catch debris in a similar manner to the rear bumper.

This photograph reveals a rather "artistic" repair found during a bare metal repaint. The sill behind the right rear tire is shown.

Rotted front floor pans are common, particularly on the left (exhaust) side. This is the result of exhaust heat and clogged duct drains. (See Section 8-1.)

Bare metal repaints reveal all manner of hidden accident damage. The best you can do is to carefully evaluate the paint to see if it is a uniform color throughout, then sight along the panels to look for waviness and orange peeled paint. Using a magnet to detect Bondo® is not reliable on these cars. The factory paint job is very thick which will give false results. In addition, the hood, trunk and door skins are made from aluminum and thus non-magnetic.

This photo reveals a large diagonal application of Bondo starting at the lower right hand corner of the fuel filler door and reaching forward above the wheel arch. Although not the case here, there is nothing inherently wrong with Bondo when applied correctly in very thin layers. The Bondo was ultimately removed from this fender and it was hammered out properly.

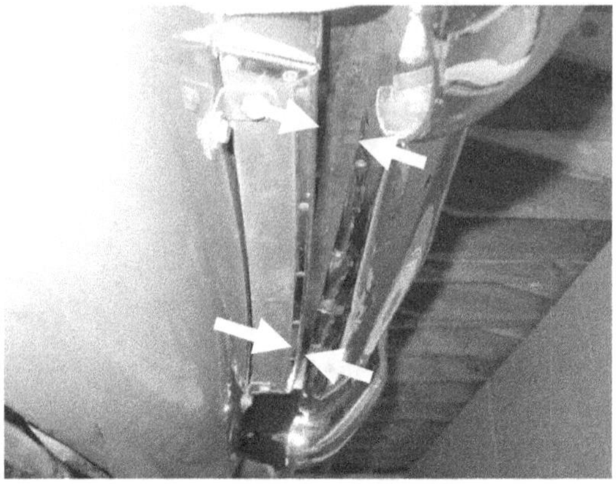

You can evaluate the bumpers by sighting along the body and trim. This is actually an insurance photo that I submitted after hitting a retread on the way back from the 2005 RROC National Meet in Greenwich, CT.

In the rear, the spring perches are attached to the inboard side of each trailing arm.

The lower end of the spring is mounted on a canvas seat in the perch. This collects moisture and rots out the perch. If the rust has made it to the outside of the cup, an expensive repair bill awaits.

The power steering cooler behind the grille will ultimately begin to leak around the flexible couplings. This takes less than an hour and a few dollars to rectify. A leaking cooler is an indication that there may be other deferred maintenance items lurking.

The "rat trap" is an enclosed box of brake linkages found underneath the driver's seat on the outside of the car. Leaks from the rat trap are not a good thing. With the rat trap cover removed, this "Deceleration Conscious Pressure Limiting Valve," or G-valve and master cylinder appear to have been leaking for some time.

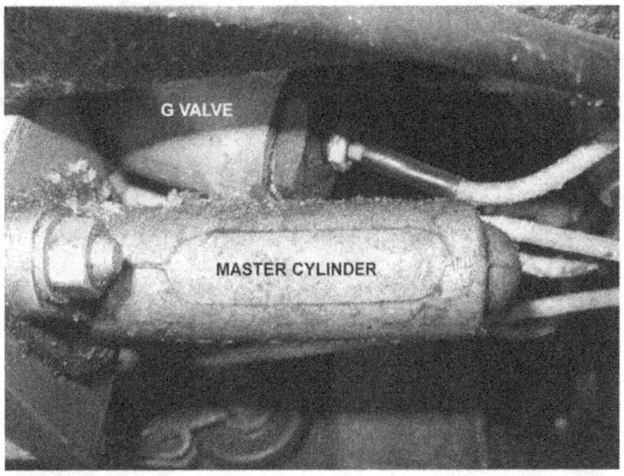

This master cylinder has blown its seal and is leaking through the protective boot. Refurbishment of these items is labor intensive, although the parts are less than $100.

This rear leveling valve left a small puddle on the garage floor. A rebuild kit is inexpensive, but servicing it is moderately labor intensive.

Shadows that sag in the rear can do so for a variety of reasons, which can run the gamut in terms of expense. Make sure you understand what you're getting yourself into or make sure it's fixed before you buy. As a rough rule of thumb, the height of the sill to the ground should be the same from front to back.

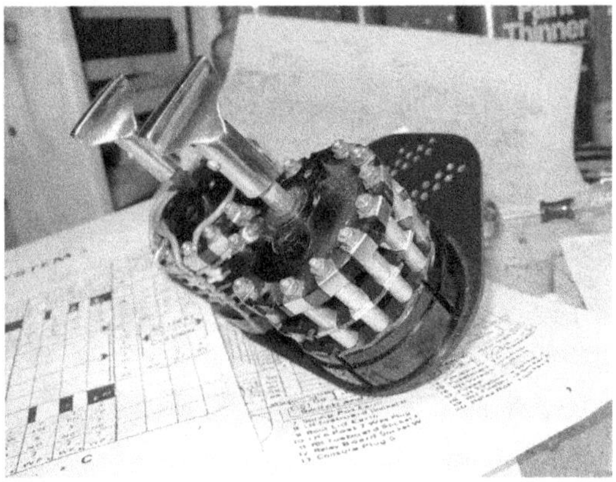

The control unit for Shadow I air conditioning is mechanically complex. Make sure all the flaps work and the A/C functions correctly. This can be very labor intensive to rectify.

The Shadow II control unit is fully electronic, which makes it more reliable yet harder to diagnose.

This is one of the contact disks removed from the unit in the previous photo. They tend to fracture with age. Getting to this level of disassembly took the better part of a day.

Failures in the Shadow II control unit are *usually* relay related and easy to fix once diagnosed.

Don't let bad looking wood trim pieces scare you off. As long as the underlying veneer is in good shape, anyone with patience and a steady hand can refinish them to an "as new" luster.

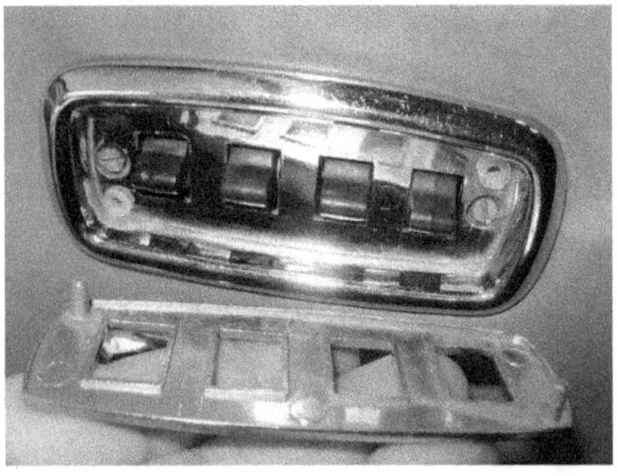

Ensure all of the interior pieces are present. These escutcheon covers in particular can go missing or have broken posts. They are expensive to replace if more than one or two are missing.

Section 1-3
Idiosyncrasies

After performing work on your Silver Shadow for awhile, you will begin to understand and finally embrace its idiosyncrasies. It goes together differently than most other cars, constantly challenging you to remember the basics of carburetion, hydraulics and electricity.

Fasteners

The Silver Shadow uses two families of fasteners: British Association (BA) and United National Fine (UNF). Fasteners larger than 3/8-in. across the flats of the nut or bolt head ("3/8-in. A/F") are UNF and easily sourced in most U.S. hardware stores. Standard inch wrenches and sockets are used and 7/16-in., 1/2-in., 9/16-in. and 3/4-in. are most popular.

British Association (BA) is a thread standard for small machine screws. Almost all small machine screws on the Shadow Series are BA with the exception of purchased parts, which use UNF hardware. As an example, the exterior lamps other than the headlamps use #6-32 screws. BA fasteners require special sockets and wrenches: 2BA, 3BA, 4BA, 6BA and 7BA are the most popular.

The original 2BA screws and nuts used on the Silver Shadow sometimes measure 11/32-in. across the flats rather than the correct 0.324-in.—you will need to stock an 11/32-in. socket and wrench as well! (See fig. 1-3-1.)

Socket Size	*Hex Size Across Flats (in.)*	*Fastener Description*	*Major Dia. (in.)*
7BA	0.172	7BA	0.0984
6BA	0.193	6BA	0.1102
4BA	0.248	4BA	0.1417
3BA	0.282	3BA	0.1614
2BA	0.324	2BA	0.1850
11/32	0.344	2BA*	0.1850
7/16	0.438	1/4-28 UNF	0.2500
1/2	0.500	5/16-24 UNF	0.3125
9/16	0.563	3/8-24 UNF	0.3750
5/8	0.625	7/16-20 UNF	0.4375
3/4	0.750	1/2-20 UNF	0.5000
7/8	0.875	9/16-18 UNF	0.5625

*common non-standard size (see text)

Figure 1-3-1 – Common fasteners on the Shadow Series

The engine block and cylinder heads are cast from aluminum alloy. For added strength and improved serviceability, Rolls-Royce uses stainless steel Heli-Coil® inserts for many of the threaded holes found in these areas. This is particularly unfortunate in the case of the thermostat housing, which succumbs to corrosion in the presence of the coolant and dissimilar metals. (The thermostat uses 1/4-28 UNF fasteners.)

One notable exception to this Heli-Coiling practice is the spark plug holes. Because of this, it is important to take extra care to avoid cross threading the plugs during installation and to limit the torque to 15-18 ft.-lbs. Always start the plug into the hole directly by hand or with a flexible rubber tube over the insulator acting as an extension.

The lug nuts that secure the wheels are made from brass and should be torqued to 45 ft.-lbs., not the much higher figures typical on other cars. This is why the tire iron (or "tommy bar") in the tool kit is so short. Unless you are a body builder it is difficult to apply more than 45 ft.-lbs. using the tool supplied with the car. Also note that the lug nuts on the left side wheels are left-hand threaded, i.e., these nuts tighten when turned counterclockwise.

Monocoque Construction

The Silver Shadow series was Rolls-Royce's first effort in monocoque or unibody construction. Whereas all previous models had a body mounted to the chassis, monocoque construction uses the body itself as the load carrying structure. This serves to reduce weight, increase stiffness (and thus improves handling) and increase the amount of available interior volume.

Monocoque construction was relatively advanced for large sedans in the mid Sixties—and particularly so considering the lack of Computer Aided Engineering tools that we take for granted today.

With this in mind, it is not surprising that there are a few weak points in the design that need to be taken into consideration, particularly when preparing to jack or tow the vehicle. (The next two sections are dedicated to each topic, respectively.)

Suspension and Brakes

The front suspension consists of two coil springs and telescopic shock absorbers with upper compliance arms and lower control arms. (Early Shadows have an upper control arm instead of the compliance arm.) The rear suspension is also a coil spring and telescopic shock design with semi-trailing arms and hydraulic height control. Early Shadows also offered hydraulic roll restriction but that quickly became problematic and all such systems have since been deactivated by dealers.

All four wheels have Girling calipers on 12.4-in. discs. There are two single piston calipers located on each front rotor and one dual piston caliper with integrated parking brake located on each rear rotor. During the 1973 model year, vented front rotors became standard.

This system stops the vehicle remarkably well, even by contemporary standards. The lines are mild steel Bundy tubes with various SAE 45 degree flare connections. They are actuated at the same pressures (up to 2500 psi) as conventional systems. The big difference between this and more typical brake systems is how the pressure is generated and modulated.

The brake pedal is connected to an enclosed box, known as the "rat trap", located beneath the driver's seat on the underside of the vehicle. It acts upon a pair of valves that route pressure to the brakes from a corresponding pair of spherical hydraulic pressure accumulators mounted on the engine. For safety, these two circuits operate independently.

Circuit 1 operates the front caliper on each front wheel and one of the pistons on each of the rear wheels. Circuit 2 operates the rear caliper in each front wheel.

Up through chassis number 22117 on the standard saloons*, a small master cylinder, also located in the "rat trap" is used to provide brake feel and generate pressure for the remaining piston on each rear caliper. For chassis numbers 22118 and higher, brake feel is provided by a rubber

---

* See fig. 2-4-5 for the exact coachbuilt chassis numbers that incorporated this change.

puck and circuit 2 provides pressure to the remaining piston on each rear caliper.

Also located in the rat trap is a simple ball ramp device known as the "Deceleration Conscious Pressure Limiting Valve," or G-valve. The G-valve cuts off the pressure to one piston on each rear wheel during hard deceleration, thereby preventing rear wheel lockup.

The camshaft, which is located directly above the engine crankshaft, has two extra lobes. These are used to drive a pair of reciprocating hydraulic pumps which are very similar to diesel injector pumps of the period. Each pump draws Rolls-Royce specified Castrol RR363 hydraulic brake fluid from its own compartment in a reservoir and forces it into the upper half of its respective accumulator. The accumulator has an upper and lower half separated by a rubber bladder. The lower half is filled with pressurized nitrogen.

The top half of the accumulator fills with fluid which pushes against the bladder causing the nitrogen to compress. This is an effective means of storing hydraulic pressure. In the event of an engine stall, the brakes will continue to be fully operational for a period of time. If the spheres are depleted of nitrogen, the sphere will fill completely with brake fluid. Since brake fluid is virtually incompressible it will not store any energy and allow no measure of safety in the event of a stall.

Atop each of the accumulator spheres, a pressure regulator can be found. This diverts brake fluid back to the reservoir once the accumulator is fully pressurized. There is also a pressure switch which illuminates warning lamps on the instrument panel when accumulator pressures become dangerously low.

Engine

The assignment for Rolls-Royce engineer A.J. Phillips and his team was to design an engine with 50% more power than the outgoing in-line 4.9L six cylinder with a level of smoothness and silence not previously achieved, without an increase in weight or cost. The result was a splendid mix of competitive benchmarking, aircraft engine metallurgy, and evolution from previous designs. The engine went into series production in 1959 in 6.23L V8 form with the introduction of the Silver Cloud II. A.J. Phillips has said that during final development, the engine passed a 400 hr. dynamometer test at full throttle and 4,200 RPM without a failure of any sort.*

When the Silver Shadow was introduced in the fall of 1965, the engine continued on essentially unchanged except for a reorientation of the upper intake system and a redesign of the cylinder heads. In 1970, the stroke was lengthened to achieve the 6.75L displacement that has lived on (in modified form) to power the present day Bentley Arnage series.

The engine block is aluminum with a wet liner design consistent with WWII aircraft engine practice. The preference would have been to use a thin wall iron casting technique pioneered by GM, which offers benefits in noise, cost and weight versus the aluminum design, but this technology was not available to Rolls-Royce at the time.

The camshaft is gear driven for low noise, following a Rolls-Royce practice that dates all the way back to the Silver Ghost era. Two overhead valves per cylinder are actuated by pushrods and hydraulic valve lifters. Valve seats are hardened and not sensitive to unleaded gasoline regardless of vintage. However, early cars with 9:1 compression ratio do prefer a steady diet of high octane fuel.

The combustion chambers are of a wedge design. Hemispheric or "hemi" heads were evaluated, but the flame propagation was too slow. Looking back,

---

* A.J. Phillips, *The Design History of a V8 Engine*, 1961-62.

this was a fortunate choice with U.S. emissions regulations around the corner, since true "hemi" heads have issues with incomplete combustion. If "hemi" heads had been chosen, a costly redesign would have been necessary in the early Seventies, adding to the barely surmountable challenge of keeping the car emissions legal in the U.S. market.

Carburetor choice followed post-WWII Rolls-Royce design practice with the fitment of twin Skinner's Union (SU) constant velocity sidedraft units. American made downdraft 4-bbl. units were evaluated, but the decision was made to keep the supply chain manageable on this critical component by selecting the domestically available SU. The Camargue did use a 4-bbl. downdraft style but this decision was made ten years later and on much smaller volumes.

The choke is a single unit separate from the carburetors that draws a metered flow of clean intake air past the left hand exhaust manifold via a small "stove pipe" to warm a bi-metallic thermocoil mechanism and open the choke plate automatically as the engine warms.

The firing order is rather unconventional but was chosen to reduce engine sensitivity to carburetor balance. As a result, however, it was difficult to tune the intake for peak power. When the bottom end of the engine was redesigned for the 1987-1/2 20,000 series Silver Spirits, the decision was made to go to fuel injection across the board. This allowed the firing order to be changed in conjunction with an intake manifold redesign. The result was a noticeable improvement in power while maintaining ultra low vibration levels at idle. The only casualty was the subtle yet distinct sound of the earlier V8 firing order.

Given that this 412 cid engine develops perhaps an estimated 200-220 bhp at 4,200 RPM, the selection of a forged steel crankshaft seems like overkill. When maintained properly, 150,000 miles between major overhauls is not uncommon. Improper maintenance is almost always the cause of things gone awry. Problems usually begin with the cooling system.

Aluminum alloy is much more sensitive to coolant condition than cast iron. As a result, coolant left in the system longer than one or two years will begin to attack the metal and the resultant sludge finds its way into the radiator core, which in turn reduces its effectiveness. This situation builds until an overheating condition occurs, at which point the block has a tendency to shift dimensionally and allow the main bearing bolts to relax, which in turn leads to chronically low oil pressure, ultimately leading to catastrophic engine failure.

Weakener

The SU carburetors do an admirable job of providing a reasonably accurate air fuel ratio under a wide variety of operating conditions. In order to improve cruising efficiency further, Rolls-Royce engineers devised a system that consists of a small orifice on the "B" side (left) carburetor connected to the (sealed) float chambers on both carburetors. At part throttle cruising speeds, the orifice draws a small vacuum (~2-in. H_2O) that slightly reduces the amount of fuel passing through the main jet on each carburetor, thus leaning out, or weakening the air fuel ratio and improving cruising efficiency.

Of course, weakening the mixture while the intake air is cold would cause driveability problems, so there is a bi-metallic switch present just behind the choke plate to detect this and disable the weakener system if this is the case.

With this system in place, additional functionality is provided through the inclusion of a solenoid that turns on and off with the ignition key. With the key off, the solenoid connects full engine vacuum to the weakener system completely eliminating the differential pressure between the float chamber and intake manifold, thus completely cutting fuel to the engine. This is called the anti 'run-on' solenoid and prevents the engine from dieseling.

Transmission and Rear Differential

As on the Silver Cloud, early home market Silver Shadows were fitted with a 4-speed Rolls-Royce built, GM licensed Hydramatic transmission. All other cars came equipped with an Allison built, 3-speed Turbo-Hydramatic 400. Second only to the Chrysler TorqueFlite A-727 in indestructibility, the TH400 is one of the most prolific transmissions of all time. It can be serviced economically just about anywhere with the possible exception of Antarctica.

Regardless of which transmission is installed, there are a few extras—most notably electric range selection. There is what looks like a large windshield wiper motor mounted to the left side of the transmission and connected to the range selector arm via a four bar linkage. The lever on the steering column is merely an elaborate electric switching device which signals the electric motor to select the appropriate range.

A fortunate byproduct of the electric range selection is the ability to disable the transmission by removing a gearshift cut-out on the fuse board. This is a very useful safety measure when diagnosing problems with the engine running.

The 3.08:1 ratio Rolls-Royce built rear differential is fed by a single piece drive shaft. The rear halfshafts use ball and trunnion joints adjacent to the differential and Hardy Spicer® joints on the wheel end. The trunnion joint is sealed and tends to leak *into* the differential case, thus it can become dry and unlubricated without providing any external evidence. (See Section 2-5.)

Engine Accessories

After WWII, Rolls-Royce adopted a practice of outsourcing standard components to reduce development costs at no expense to quality. Consequently, almost all of the major engine accessories are sourced from the same supplier base used by other automotive concerns of the period: Lucas, Eaton, GM, SU and others. This practice is good news for you in terms of price and availability of replacement parts.

The air conditioning system is comprised of a GM Harrison A-6 compressor with a suction throttling valve by Singer (the sewing machine company) and Lucas electric servos actuating all of the ductwork flaps. The automatic controls found in the Silver Shadow II, however, were an industry first and are a proprietary Rolls-Royce design.

Early power steering pumps were manufactured by Hoburn Eaton, while later units were made by GM Saginaw until the introduction of rack and pinion seering in the Silver Shadow II.

The coil and distributor were provided by Lucas through various iterations as was the starter. Varieties of alternators (and even a generator) were sourced to Lucas and General Motors. Most of the air pump hardware can also be traced to GM ACDelco designs.

Section 1-4
Jacking

Jacking deserves its own section because there are many easy ways to cause damage to your Silver Shadow or Bentley T. *First, regardless of the jacking method employed, make sure that the wheels that will remain on the ground are chocked front and back.* Second, you can never go wrong using the Rolls-Royce supplied jack in one of the two specified jacking locations. These are located on either rocker panel halfway between the front and rear wheels.

Beyond this, it is important to understand the vulnerabilities of your automobile. The most expensive mistake is to attempt to lift the car by the rear subframe. This is the structural member to which the differential is attached. It looks like a bridge truss, but the structure will readily deform under the point load presented by a floor or bottle jack. If the subframe is damaged, replacement parts costs alone will exceed $2,000. Deformation of this vital component is an important item to check when evaluating a new car for purchase as mentioned in Section 1-1.

The basic monocoque structure is very robust, and can be safely lifted using a four point hoist on the structural portion of the sills behind the front wheel arches and in front of the rear wheel arches. (See fig. 1-4-1.) Lifting at any one of these points singly with a floor or bottle jack should be avoided because it may twist the structure just enough to crack the paint around leaded seams, particularly near the base of the A-pillars.

For lift work, the safest option is to use a lift with two runners that you simply drive onto. Ramps are also useful. Remember to use a set rated for full size pickups or SUVs; your baby weighs almost 2-1/2 tons.

In the front of the car, a floor jack can be

Fig.1-4-1 – If you absolutely need to lift the car by the body, use these points on both sides. (See text.)

used on the lowest point of either lower control arm. Alternatively, you can raise the entire front of the vehicle with a floor jack and a suitable block of wood placed where the two control arms meet in front of the engine. (See figs. 1-4-2 and 1-4-3.)

Be careful to avoid crushing the exhaust system or the transmission cooler lines exiting the bottom of the radiator.

To support the engine while replacing the engine mounts, you can use a floor jack in conjunction with a 3/4-in. thick square of plywood to lift on the stamped steel oil pan. To support the radiator in preparation for removal, you can gently support it from the bottom with a bottle jack. (Under no circumstances should either of these methods be employed to lift the front of the car.)

I have seen mechanics raise a rear wheel using a bottle jack on the rear trailing arm as shown in fig. 1-4-4. If you are a professional, this practice will not damage the trailing arm.

If you are not a professional, don't do it! While raising the rear wheel and trailing arm with the bottle jack, the bottom surface of the trailing arm will change orientation as it moves closer to the chassis. This may cause the bottle jack to lose contact with the trailing arm, which will allow the wheel, trailing arm, and the rest of the car to come down with potentially lethal results.

Fig.1-4-2 – Location for floor jack on a lower control arm.

Figure 1-4-3 – Floor jack location where control arms meet.

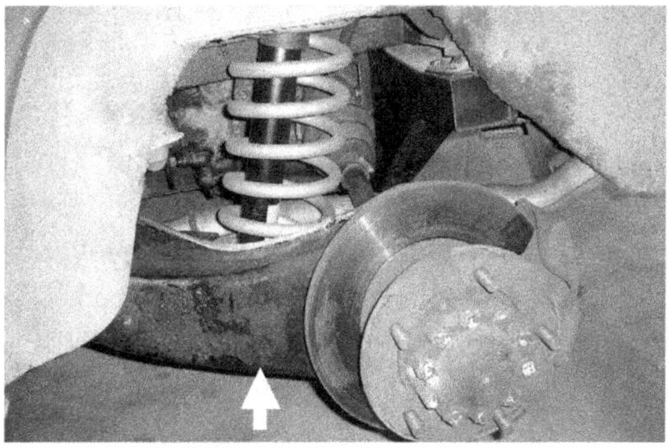

Figure 1-4-4 – Professionals only: jacking location at rear trailing arm. (See text.)

Section 1-5
Towing

You can spend more money fixing the damage to your car *after* being towed incorrectly than fixing whatever was wrong prior to the tow truck's arrival. There are plenty of wrong ways to secure the car for towing including the rear subframe and the positive battery cable. The proper procedure:

Figure 1-5-1 – Front lower control arm.

1. Insist on a flatbed. Period. If the car needs to be winched on to the bed, attach a cable to one or both of the lower control arms in the front. (See fig. 1-5-1.)
2. Chrome bumper Shadows are often equipped with eyehooks below each overrider on the rear bumper. (See fig. 1-5-2.) These can be employed as an alternative winch location if traffic permits the flatbed to load the car rear first. *Be careful to avoid the positive battery cable that exits the trunk behind the left rear wheel well.*

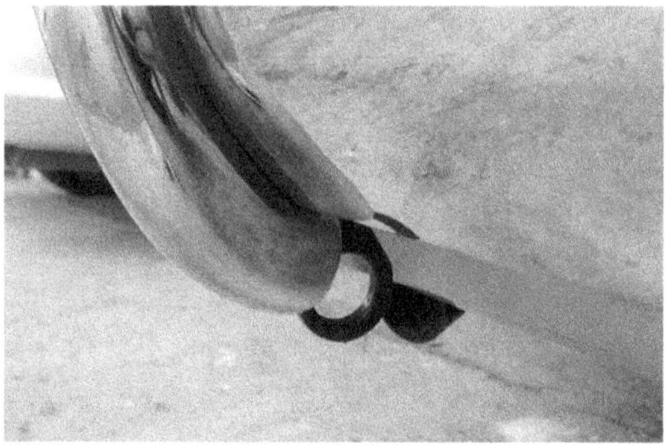

Figure 1-5-2 – Eyehook at bottom of rear bumper overrider.

3. Use nylon tie downs to secure each wheel to the truck. (See fig. 1-5-3.) You can carry four of them with you. Crewe parts even markets an "official" set.
4. When removing the vehicle from the flatbed, use the winch. *Remember: a non-running Shadow may not have enough pressure in the hydraulic accumulators to stop.* Attach the tow truck's winch cable as in 1 or 2 above.

Figure 1-5-3 – Nylon wheel tie down.

Section 1-6
Tools

As discussed in Section 1-3, the larger fasteners are standard inch type and those less than 3/8-in. across the flats of the nut or bolt head (3/8-in. A/F) are British Association (BA). In addition, special tools are often needed to negotiate the extraordinarily tight engine package. This is particularly the case on later engines with emissions equipment.

<u>Recommended Basics</u>

- SAE socket set, 1/4-in., and 3/8-in. drive. Six point, standard and deep well, 11/32-in. through 7/8-in. There are no metric fasteners on this car.
- British Association 1/4-in. drive socket set, ranging from 2BA through 7BA.
- British Association open end wrench set, ranging from 2BA through 7BA.
- Phillips and flat head screw drivers, various sizes and lengths.
- SAE combination wrench set, 11/32-in. through 7/8-in. Ratcheting box end wrenches also come in handy.
- SAE crows feet wrenches, 11/32-in. through 7/8-in.
- Stubby wrenches, 3/4-in. in particular.
- Extension bars in various lengths, 1/4-in., and 3/8-in. drive.
- Universal joints, 1/4-in., and 3/8-in. drive.
- Wobble bars in various lengths. 3/8-in. (Similar to extension bars, but the wobble feature allows for slight off-axis use.)
- Breaker bar, 3/8-in. drive.
- Torque wrench, 0-50 ft.-lbs.
- Multimeter that includes a diode test function.
- O-ring pick.
- Magnet on a stick.
- Mirror on a stick.
- Flexible "claw" or "pickle grabber."
- Soft fender protectors.
- Inch size Allen keys, 0.050-in. and up.
- Large capacity oil pan, 9-10 qts.
- Large capacity coolant pan, 5-6 gal.
- Mini ratchet set from Griot's Garage: p/n 10247. (See fig. 1-6-1.)
- Floor jack.
- Bottle jack.
- Heavy duty jack stands.
- Wood or polyurethane blocks or pucks to distribute load and protect underbody when jacking.

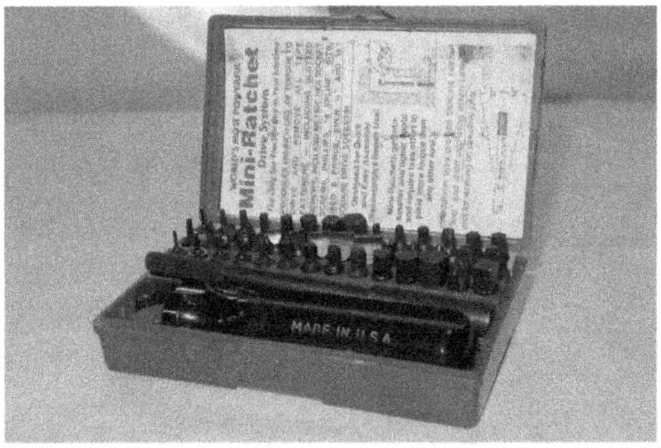

Figure 1-6-1 – Mini-ratchet set.

For Hydraulic and Brake work

- 1-in. deep well socket for outlet screens.
- 1-3/8-in. socket for G-valve.
- #1 "EZ-out" for reservoir sight glass screws.
- #2 "EZ-out" for G-valve.
- Set of inch sized flare nut wrenches.
- Pressure or vacuum bleeder (for Shadow I models with master cylinder—see Section 2-4).

Figure 1-6-2 – IR Thermometer.

For Engine Tune-Up

- Timing light.
- Engine analyzer with dwell and tachometer settings.
- Manometer. (See Section 9-4.)
- Gunson® Colortune™. (See Section 3-5.)

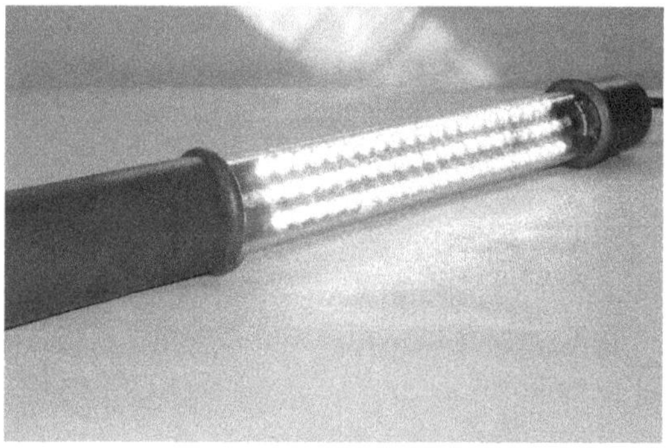

Figure 1-6-3 – Rechargeable LED work light.

Consumables

- Anti-seize paste.
- High temperature RTV Silicone.
- Permatex® No. 2 gasket sealant (for valve cover gaskets).
- Gasket remover spray.
- Penetrating Oil, e.g., PB Blaster®.
- Dielectric grease.
- Shell Rotella-T® 15W-40 Motor Oil.
- Castrol™ RR363 Brake Fluid (except 50,000 series Corniche and Camargue).
- Mobil 1® Synthetic Grease.
- Mobil 1 75W-90 Synthetic Gear Lube.
- Loctite® or Permatex "Red"—high strength, for fasteners 3/8-in. and larger.
- Loctite or Permatex "Green"—low strength, for fasteners up to 1/2-in.

Other

- 1-1/16-in. and 1-1/2-in. open end wrenches for A/C work.
- Two 1-in. open end wrenches for air pump one way valves.
- Air ratchet, 1/4-in. drive (and suitable air compressor).
- Heat gun and shrink tubing.
- IR Thermometer, for evaluating A/C and cooling system, finding cylinder misfires. (See fig. 1-6-2.)
- 1/4-28 Heli-Coil thread repair kit (p/n 5402-4), for thermostat housing.
- Rechargeable LED work light. (See fig. 1-6-3.)

Chapter 2
Basic Maintenance

Section 2-1
Change Oil and Oil Filter

This is one of the simplest things to do for your motor car and is part of the holy trinity of Shadow maintenance: fresh oil, fresh coolant and fresh brake fluid. The Handbook (Owner's Manual) states that you should change your oil every 3,000 miles and your oil and filter every 6,000 miles. If you drive less than 3,000 miles per year, plan on changing your oil and filter every 12 months, preferably in autumn prior to storing the car for the winter.

It seems that everyone has an opinion on oil. Rest assured *changing* it is far more important than the type of oil that you ultimately select. For non-catalyst cars, I like the commercial 15W-40 mixed fleet truck oils that you can buy in one and five gallon containers such as Mobil Delvac® and Shell Rotella-T.

Back in the Nineties, Patrick Bedard wrote a column in *Car and Driver* about the right motor oil for 20-50 year old "keeper cars." He mused that custom blended oil for old cars would be high detergent and have a greater complement of anti-wear and anti-corrosion additives because they spend so much of their time sitting as compared to a daily driver. His research revealed that one of the best anti-wear and anti-oxidation additives around is zinc dialkyl dithio phosphate (ZDDP). Unfortunately, it has been all but eliminated in modern passenger car oils because it fouls catalysts.

Fortunately, heavy duty mixed fleet engine oil comes very close to fitting the bill. It is designed to neutralize the acidic compounds generated by highly sulfuric diesel fuels and has 80% more ZDDP than an oil designed for a contemporary passenger car. (Heavy duty trucks don't have catalysts.) And, it's only $8 per gallon!

The price of an official Crewe cartridge type oil filter for the Shadow I has escalated quite a bit at the time of this writing. There are substitutes available on the internet. Make sure that they come with all of the extra washers and seals supplied with the official part. The spin-on style oil filter for the Shadow II is priced more reasonably. In any event, you will need two gallons plus one additional quart of oil.

Obviously, if you need to drain two plus gallons of oil, you will need something larger than a standard five quart drain pan to hold it. The oil comes out *fast,* so one large pan is better than trying to swap pans in the middle of the operation.

Changing the oil after a nice long highway cruise is desirable because hot oil leaves the engine more readily. That also means you have to deal with hot engine parts and hot oil. Be careful.

To reduce risk of injury or death while working underneath the vehicle, chock the wheels and disable the transmission by removing the gear shift cut out from the fuse board.

Changing the oil with the car on a lift is a luxury. For the rest of us, it's more than likely that it is necessary to jack the car up to get the oil pan under the sump. Jacking from the left control arm works best as you will not need to change jacking locations when it comes time to change the filter. (See fig. 2-1-1. Refer to Section 1-4 for more information on jacking.)

Locate the oil drain plug key in your tool kit. Quite often these get misplaced over time. Fortunately, authentic second hand reproduction drain plug keys are fairly easy to locate. (See fig. 2-1-2.)

With the drain pan in position, insert the large end of the key in a 13/16-in. socket with a suitable ratchet and remove the drain plug and washer. (See fig. 2-1-3.) Watch out for that hot exhaust pipe! The washer often sticks to the oil pan. Make sure it has been removed to avoid a double washer situation.

Once the oil has drained completely, clean and replace the drain plug and washer. Tighten judiciously: the pan is made of stamped sheet metal, not machined from a block of Kryptonite.

Figure 2-1-1 – Insert jack from front underneath control arm.

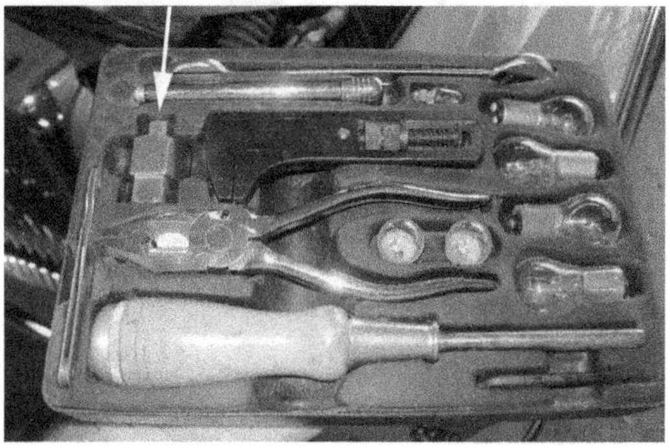

Figure 2-1-2 – Tool kit with oil drain plug key.

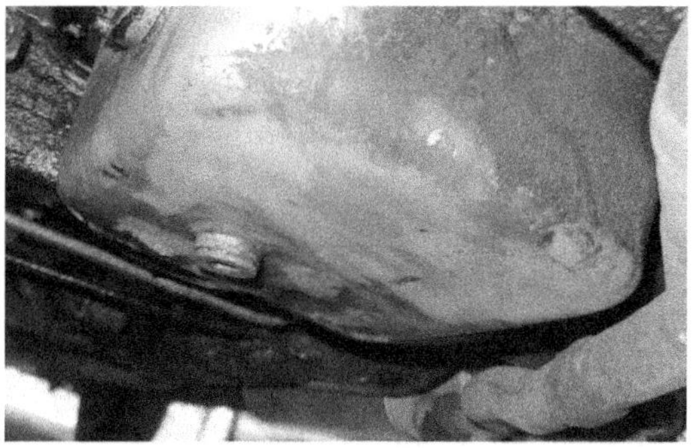

Fig. 2-1-3 – Drain plug location on sump.

Shadow I Oil Filter

The oil filter is located on the left side of the engine in front of the two hydraulic accumulators and canted at a 45 degree angle. Place the oil pan underneath the filter housing and loosen the bolt on the end with a 7/8-in. socket. This bolt has fine threads so it will take quite a few turns to remove completely. Support the housing as it loosens so that the threads are not damaged as it comes away from the engine. Again, be careful not to burn yourself on exhaust components.

Using an o-ring pick, remove the mating rubber sealing ring from the recess where the filter housing mates to the engine. Be careful not to scratch the aluminum.

Disassemble all of the items contained in the filter housing. Clean the inside of the housing with a lint-free rag and reassemble the oil filter using the items found in the kit. (See Fig 2-1-4.) Note that there will more than likely be an extra large sealing ring that came with your new oil filter. It is only used on Silver Cloud V8s and can be discarded.

Make certain that the filter housing lip and the recess for the sealing ring are clean. Lubricate the sealing ring that you *do* need (item 8 in fig. 2-1-4) with clean engine oil and carefully install it all the way into its recess on the engine. Holding the renewed filter assembly, carefully turn the retaining bolt clockwise a few turns to avoid stripping

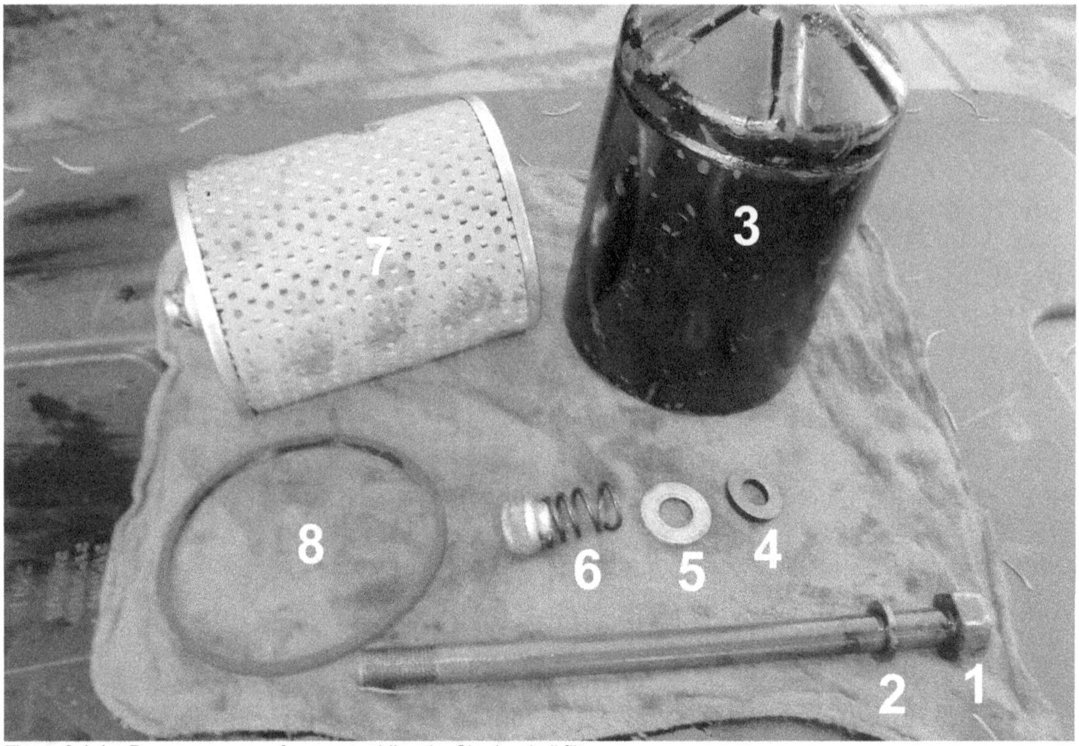

Figure 2-1-4 – Proper sequence for reassembling the Shadow I oil filter.

Fig. 2-1-5 – Shadow II oil filter shown with engine removed.

the threads. Once it is engaged sufficiently, push the housing into the sealing ring recess. Tighten the bolt with the other hand while holding the housing firmly in the recess. It is very easy to get a poor seal if you do not pay attention to this step.

Shadow II Oil Filter

With a drain pan underneath, use a filter wrench to remove the old oil filter. Clean the oil filter sealing surface on the engine. Lubricate the seal on the new oil filter with a few drops of clean engine oil and thread it on. Tighten it ½ to ¾ turns only with the filter wrench.

All Models

Double check that you installed the drain plug, then refill the engine with 7 quarts of oil and check the dipstick. Add additional oil to bring it to halfway between the max. and min. marks on the dipstick.

Start the engine. Allow it to idle—do not rev it! Listen first for squirting noises in the vicinity of the oil filter. If you hear that, turn the engine off immediately. You have a mess to clean up because the oil filter is not seated properly. (This only happens to the Shadow I owners.) The oil light should extinguish within five seconds or so. If it does not extinguish, quickly turn the engine off. Did you really put oil in the engine?

Let the engine sit for awhile, put your tools away, etc.—enough time for the oil to drain back into the sump. Check the dipstick again and add enough to get to the halfway mark. There is a difference of two quarts between the max. and min. marks.

Section 2-2
Change Coolant and Thermostat

Maintaining the cooling system in top condition is of paramount importance on the Silver Shadow. When a Shadow engine overheats, the main bearing bolts tend to loosen, causing an out of tolerance condition on one or more of the crank bearings, which in turn leads to dangerously low oil pressure and catastrophic damage—a very expensive issue easily avoided with consistent care.

Because the heads and the block are aluminum, corrosion can become an issue if the same coolant is left in the engine for more than two years. This leads to constricted passageways in the engine and a buildup of residue in the radiator core which degrades cooling efficiency over time.

The thermostat is on the same schedule because spending $67 on a new thermostat every two years is great insurance against the $20,000 that you will need for a complete engine rebuild if the part fails.

Fortunately for me, the previous owner of my Silver Shadow subscribed to this philosophy and went so far as to install an aircraft style combination oil pressure and water temperature gauge in place of the clock. This offers peace of mind by warning of detrimental conditions in the engine in advance of the warning lamps illuminating, by which time the damage is already done. (See fig. 2-2-1.)

Prior to starting work on the car, place the new thermostat (PG58252) in a sauce pan with enough water to cover it. Heat the water on a stove until it comes close to boiling. The thermostat should open at approximately 185°F. (See fig. 2-2-2.) When the thermostat is placed as shown, the center part moves downward to open.

Figure 2-2-1 – Aircraft style combination gauge.

Figure 2-2-2 – Testing the new thermostat.

Note: Anytime the cooling system is opened up, it is important that the engine be cool. Once the coolant is hot it is also under pressure and will forcibly spray scalding liquid all over you.

Starting with a cold engine and a large pan positioned under the lower radiator hose, remove the radiator cap, then drain the coolant by releasing the hose clamp where the lower radiator hose attaches to the radiator. Pull the connection loose and the coolant will drain out rapidly. In my opinion, the nearby drain tap works far too slowly.

If you have recently changed the coolant and only need to change the thermostat, drain approximately two quarts by opening the drain tap on the left bottom of the radiator

For Shadows with chassis numbers 6000 and higher, part of the air pump system needs to be removed prior to accessing the thermostat. First, some nomenclature: the *air pump* is driven by a single belt and sits adjacent to the air conditioning compressor. It has a hose exiting from the side or rear connected to a cast *four way connection.* Two of the hoses exit left and right and are connected to *check valves (UE74483)* which lead to the exhaust manifolds via the *air manifold assembly* (a.k.a. "antlers"). The remaining hose from the four way connection is connected, via a trombone slide shaped pipe and sometimes a pressure regulator, to the *gulp valve.* (See fig. 2-2-3.)

The elements of the air pump system varied over the duration of the model run. The details in this section are specific to chassis numbers from approximately 14300 to 20700, but the general idea is the same through the end of the Shadow II model run.

To gain access to the thermostat, remove the trombone slide shaped pipe and the hose that connects it to the four way connection after loosening the hose clamps at each end. Release the four way connection from its bracket by removing the two 2BA screws.

Release the hose clamps at each check valve. Release the hose clamp at the air pump itself. Lift out the four way connection together with the hoses leading to the check valves and air pump.

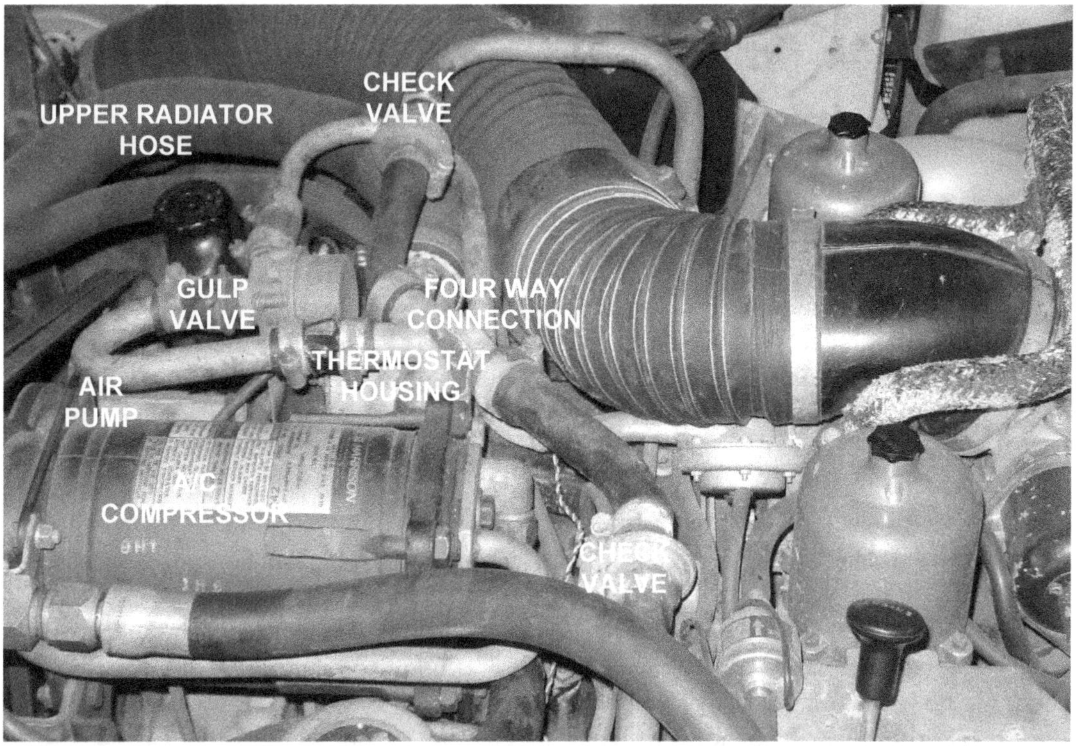

Figure 2-2-3 – Component identification.

Remove the two slotted screws that hold the gulp valve to its bracket. (See fig. 2-2-4.) Twist the gulp valve so that the thermostat housing is accessible. Disconnect the upper radiator hose from the thermostat housing after loosening the associated hose clamp.

Loosen each of the 7/16-in. A/F screws on the thermostat housing and remove them. ("A/F" means across the flats of the nut or bolt head.) After you have broken the screws loose by hand, an air ratchet can speed up the removal process as the screws are fine threaded and relatively long. Disconnect the temperature sensor wire and remove the thermostat housing top along with the upper radiator hose and set aside. Remove the gulp valve bracket, the top of the thermostat housing, and the thermostat gasket. (See fig. 2-2-5.)

Usually the thermostat is lightly wedged in its pocket. Lightly grip it with a pair of pliers and pull it straight out. Ensure that both gasket surfaces are clean, then insert the new thermostat; install a new gasket (UE73635); replace the top of the thermostat housing, gulp valve bracket, screws and washers; align and tighten. Be careful not to over tighten—the screws need only to be tight enough to compress the gasket. Use antiseize paste on the screws. You will thank yourself the next time you tackle this job.

The screws and Heli-Coils installed in the housing are susceptible to galvanic corrosion especially in the presence of the coolant and it is fairly common for one of the Heli-Coils to fail and strip out.

If this occurs, buy a 1/4-28 Heli-Coil brand repair kit, which comes with the correct inserts, installation tool, oversize tap and drill bit. Follow the instructions to the letter. You will be amazed at how quickly the drill bit will remove metal so be very careful to hold the drill vertically and mark the correct drilling depth on the bit with a piece of masking tape. Also take great care to prevent any removed material from falling into the recess where the thermostat

Figure 2-2-4 – Remove gulp valve from bracket.

Figure 2-2-5 – Old thermostat visible.

goes. (Heli-Coil kits are available from McMaster-Carr or a *good* local hardware store.)

You may want to take a hard look at your air pump hardware before reinstalling it. Now is a good time to change hoses and check valves. If this is the case, refer to Section 4-5.

Reattach the upper radiator hose, reconnect the temperature sensor and twist the gulp valve back into position. Ensure that the small vacuum line is still connected to the gulp valve.

Return the four way connection with check valve and air pump hoses to its correct position. Connect the air pump hose first, followed by the check valve connections. Finally, reattach the four way connection and gulp valve to their respective brackets. Ensure all of the hose clamps are snug.

Reconnect and tighten the lower radiator hose if you have not done so already. Reattaching it is pesky; smear some rubber friendly lubricant such as KY onto the inner diameter of the hose to help with this. Refill the engine with coolant. Use a conventional coolant such as Prestone® in a 50:50 mix with *distilled water**. Start the engine and check for water and air leaks.

After running the car through a few drive cycles and allowing it to cool, recheck the coolant level.

---

* Many European motorists will tell you to steer clear away from the traditional coolants because they contain phosphates. Phosphates tend to create deposits in the presence of high mineral content water. If you use distilled water, which costs less than $1.00 gallon, by the way, this becomes a non-issue. I would recommend staying away from the orange-colored extended life coolants with organic acid technology (OAT) which seem to have been problematic. There are also hybrid organic acid (G 05 or HOAT) coolants that Mercedes has used with good results since the 1980s. This may be deemed the way to go for vintage aluminum engines in the future.

Section 2-3
Grease Fittings

There are six grease fittings on the Shadow I steering linkage: one at each track rod/wheel end joint, one at each track rod/cross beam joint, one at the cross beam/steering lever joint and one at the cross beam/idler lever joint. On the Shadow II, which is fitted with rack and pinion steering, there are only four grease fittings: those at each end of the two track rods.

The grease fittings have a Rolls-Royce specific shape and accept a special adapter that attaches to a standard grease gun. (See figs. 2-3-1 and 2-3-2.) The adapter is available from the RROC Club Stores. Use a high quality all purpose grease such as Mobil 1 Synthetic Grease.

Clean each joint with brake cleaner and a rag. Then, attach the adapter and grease gun to each joint and add some grease. Typically, one and a half to two shots per fitting are all that is needed. Add grease until the rubber seal is firm but not distorted. If the seals are in good shape you will not see excess grease exit the joint. Take care not to overfill the joints, or you may damage the seals. Grease the front suspension every two years or 12,000 miles.

Fig. 2-3-1 – Grease fitting on right track rod/wheel end joint.

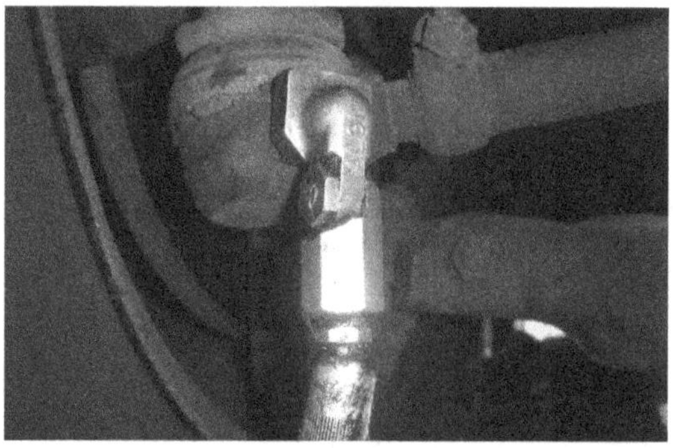

Fig. 2-3-2 – Adapter and grease gun attached to joint.

Section 2-4
Change Brake Fluid

All brake fluid is ideally incompressible for positive feel, non-reactive to preserve the integrity of the seals, and will not vaporize in the presence of heat generated by the brakes. In addition, brake fluid used in the Silver Shadow needs to lubricate the engine driven hydraulic pumps.

The correct brake fluid for all Silver Shadow and derivative models, with the exception of the 50,000 series Corniche and Camargue is Castrol RR363 which is available through your local Bentley dealer. It is similar to a conventional DOT3 type brake fluid except that it has a special additive package to deal with the constant action of the hydraulic pumps. As of this writing, RR363 prices have risen from $15 to $25 per quart. Even so, do not be tempted to substitute a conventional brake fluid!

Mineral oil is used in the 50,000 series Corniche and Camargue as well as the entire range that followed the Silver Shadow

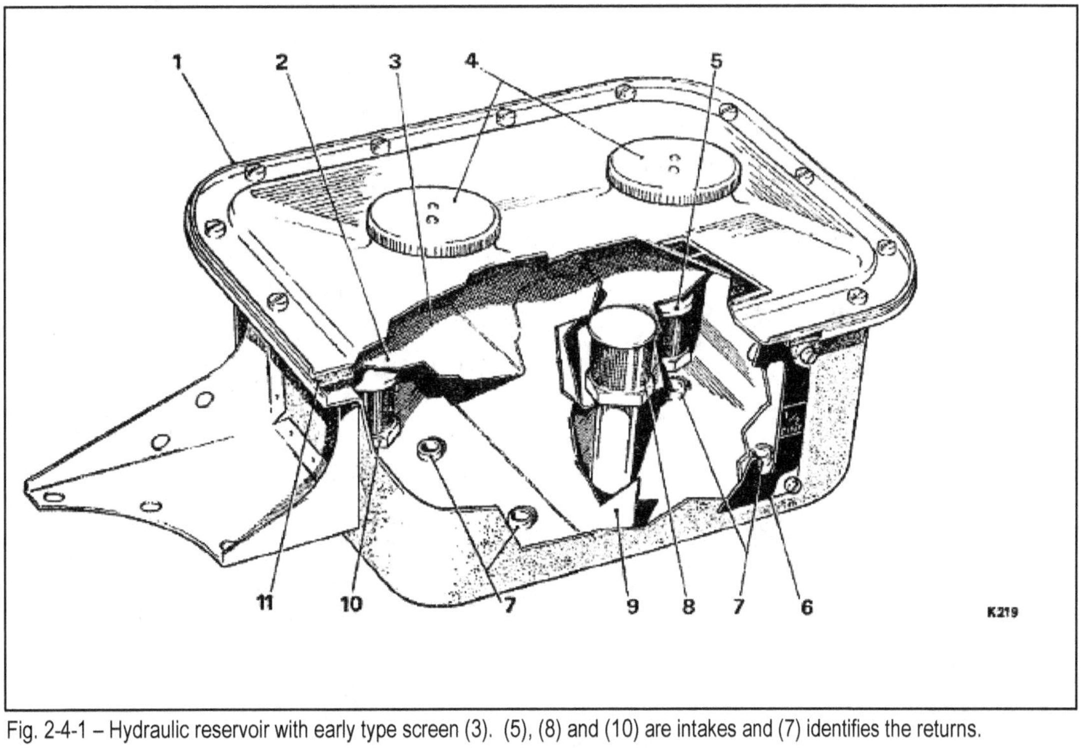

Fig. 2-4-1 – Hydraulic reservoir with early type screen (3). (5), (8) and (10) are intakes and (7) identifies the returns. (Illustration supplied courtesy of Bentley Motors Inc.)

Series in 1981. Mineral oil is generally superior to conventional brake fluid except that it will attack the seals in your Shadow Series automobile. Do not use it in place of RR363!

Five quarts of RR363 are needed to flush the brakes and hydraulics in their entirety. There are two reasons to do this. RR363 is hygroscopic, which means it absorbs moisture from the environment. Moisture laden brake fluid, whether it be RR363 or conventional, will vaporize at high temperatures much more readily. This allows the fluid to compress which decreases braking performance dramatically, especially at high speed. Moisture laden brake fluid also encourages the ferrous elements in the braking system to rust. This is why the fluid in the braking system takes on the appearance of iced tea over time.

As a rule of thumb, plan to change the fluid in the hydraulic system (brakes and height control) of your Silver Shadow every two years. Use the color of the fluid in the reservoir sight glasses as your guide. (See fig. 2-4-1.) Cars kept in arid climates may go a bit longer between changes and cars kept in very humid environments such as Florida may need more frequent service. The brake fluid ages whether or not the car is used. Fully overhauling a neglected brake system is fantastically tedious and will cost about $9-10,000 to complete at a professional repair shop. (This is why the purchase of a low mileage Shadow should be approached with great caution. Brake fluid deteriorates whether the car is used or not.)

RR363 brake fluid will remove the finish from any painted or lacquered surface, including the engine bay, exterior paint, interior wood and leather. Lay down multiple protective layers over the fender and fender apron before proceeding. And for Heaven's sake, do not wipe brake fluid from your hands onto your pants and then sit or lean on your car!

It is critically important that contamination not be introduced into the system. Many components in the system rely on very small tolerances to operate correctly and will not tolerate foreign matter of any sort.

Fig. 2-4-2 – Cork seal (deteriorated), later type fill screens.

Fig. 2-4-3 – Interior hardware removed for cleaning.

Proper service in this area includes complete evacuation and inspection of the reservoir. To begin, remove any accumulated dust or dirt from the reservoir lid, then remove all of the 2BA screws that hold the reservoir lid to the reservoir. Some of the screws thread into the reservoir lip, others have nuts and washers on the underside of lip. There should be a cork seal present underneath the lid, and for early cars, a large flat fill screen. Later cars have fill screens integrated into the baffles. Order replacements if any parts are damaged or missing. (See fig. 2-4-2.)

Using a turkey baster or similar, evacuate the fluid from the reservoir into an empty brake fluid container or Mason jar. With the majority of the fluid removed from each reservoir, inspect the three intake screens and magnetic level floats (if equipped) for damage. (See figs. 2-4-2 and 2-4-3.)

If the interior surface of the reservoir is rusty, it needs to be removed and replated or replaced with a new or serviceable second hand unit.

Carefully remove the baffle plates by releasing the 2BA nuts, bolts and washers that hold them against the dividing wall. If any of the three outlet filters are damaged, removal is straightforward. They typically have a hexagonal base that accepts a 1-in. deep well socket. Early cars use a center set screw instead.

Remove the sight glass instruction plate which is held in place by four set screws. A portion of the instruction plate (6) is shown in fig. 2-4-1.

Remove the four 5/64-in. Allen head screws to gain access to the two sight glass sealing rings. These set screws are notoriously difficult to remove. A few days prior to performing this service, you can remove the instruction plate and dribble penetrating oil on each of the four Allen screws. If the Allen key strips out the recess in one or more screws when it comes time to remove them, use a #1 EZ-out to remove the damaged fastener and replace it with a new one kept on hand for just such an emergency.

Carefully remove each sight glass from its respective retaining plate and remove the old sealing rings.

Clean the reservoir, reservoir lid, outlet filters, baffle, sight glasses and retaining plates using denatured alcohol. Allow all of the components to air dry or use compressed air to speed up the process.

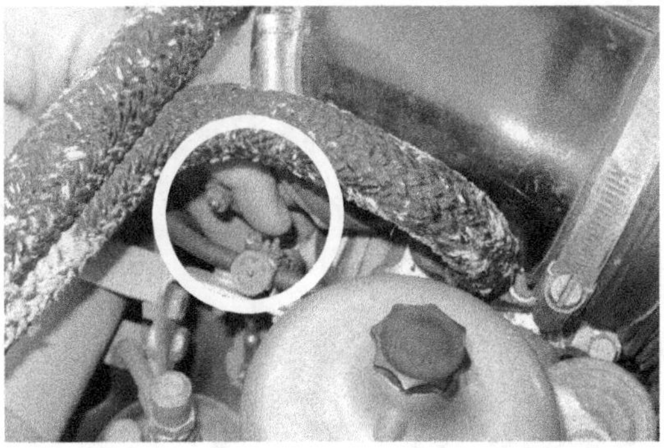

Fig. 2-4-4 – Hold choke lever down to set high idle.

When reassembling the sight glasses with new sealing rings, assemble them without the instruction plate using both the Allen screws and slot head screws. Torque each sight glass assembly down uniformly to avoid applying stress to the sight glass. Once each sight glass is installed securely, remove the slot head screws and reinstall them with the instruction plate in place.

Reinstall the baffle, screen, cork seal and lid. Refill the reservoir with fresh RR363 from new, sealed containers.

Bleeding the High Pressure System

To finish this process, all of the brake lines need to be bled. Because the brake lines do not recirculate brake fluid back to the reservoir like the height control system does, they need to be bled in order to

Chassis	Front Wheels, All Calipers	Rear Wheels, Upper Bleed	Rear Wheels, Lower Bleed
Standard: 1001-1898 Coachbuilt: 1001-1873 except 1807 and 1869	High Pressure	Master Cylinder	High Pressure
Standard: 1899-22117 Coachbuilt: 1807, 1869, and 1874-22582, except 21729 and 21998	High Pressure	High Pressure	Master Cylinder
Standard: 22118+ Coachbuilt: 21729, 21998, 22583+ LWB: 22073+	High Pressure	High Pressure	High Pressure

Fig. 2-4-5 – Pressure source for brake bleeds.

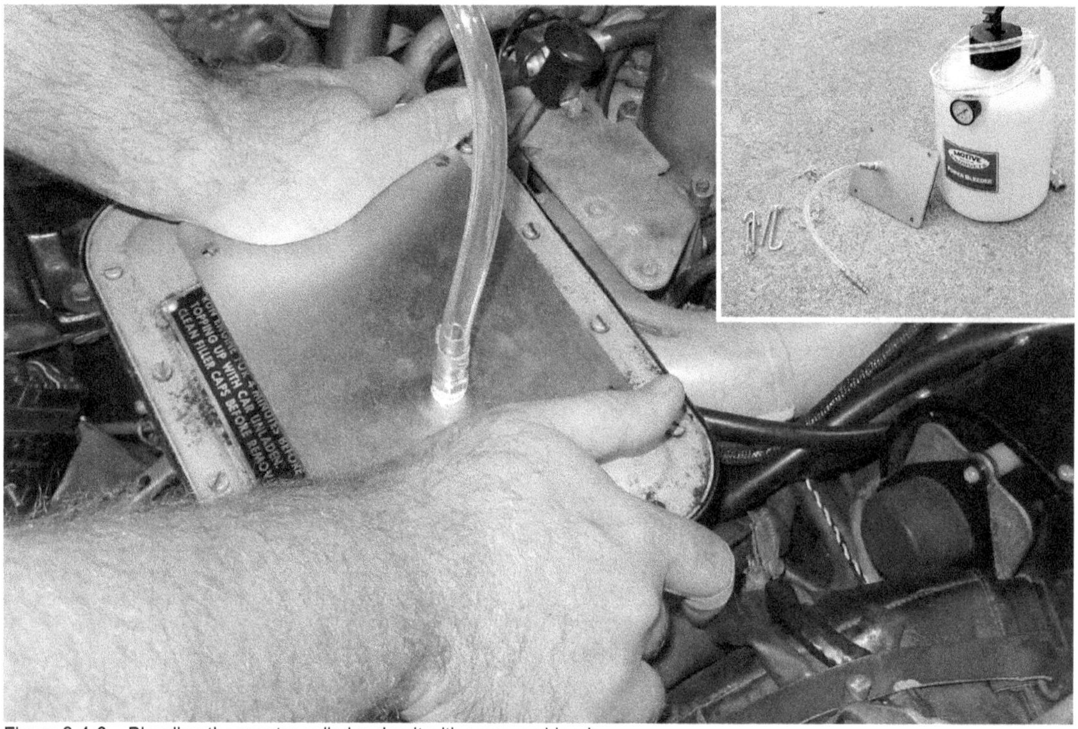

Figure 2-4-6 – Bleeding the master cylinder circuit with pressure bleeder.

remove contaminants and trapped air.

Fortunately, the task is made simpler than on other vehicles because the engine does the work of pumping the brake fluid for you, with the exception of the master cylinder circuit.

To reduce risk of injury or death while working underneath the vehicle, chock the wheels and disable the transmission by removing the gear shift cut out from the fuse board.

With the engine off, set high idle by manually holding the choke closed while pulling the throttle linkage part way open. Then allow it to close onto the high idle cam. (Figure 2-4-4 shows choke linkage held in closed position.) Start the engine and allow it to run for 2-4 minutes before proceeding.

You will need an assistant to bleed the hydraulic system. Depending on the size of person designated to work under the car, the bleed fittings can be accessed without raising the car on a lift or jacking up one side at a time using the factory supplied jack.

Start with the accumulators. On Shadow I, both accumulators are on the left side of the engine. Shadow II's have one accumulator on each side of the engine. Attach a length of clear plastic tubing to the bleed fitting on one of the accumulators with the opposite end in an empty brake fluid container or Mason jar.

Be very careful, the exhaust system runs right through this area and it is very easy to burn yourself. Bleeding the accumulators with the car raised on a lift is a safer (and easier) alternative.

Open the bleed fitting using a 7/16-in. flare nut wrench and allow fluid to flow until fresh clear fluid is flowing through the tube with no air bubbles. Repeat for the other accumulator.

While one person pushes the brake pedal halfway, bleed each of the high pressure brake caliper bleed fittings in the same manner as on the accumulators. Consult fig. 2-4-5 to ascertain which bleed fittings are high pressure and which, if any, are fed by the master cylinder circuit. Tighten all bleed fittings to 8-10 ft.-lbs. using a six point 7/16-in. deep well socket. Turn the engine off.

Bleeding the Master Cylinder Circuit

For cars so equipped, it is very difficult to bleed the master cylinder circuit by pumping the brake pedal as recommended in the Workshop Manual. Using a Motive Products™ Power Bleeder™ or similar is a significant improvement. This type of bleeder pressurizes the brake reservoir to about 10 psi and forces the fluid through the master cylinder to an open bleed nipple. Ordinarily, the bleeder can be clamped in place on all manner of vehicles, but the Shadow reservoir is too large.

Rather than going though the trouble of modifying the clamping mechanism, it is easy enough to hold it in place with the tube positioned over reservoir #2 and the rubber sealing the opening over reservoir #1 while your helper pumps up the Power Bleeder and opens each low pressure bleed nipple in turn. The two reservoirs share airspace so they both need to be sealed off. (See fig. 2-4-6.) Bleeding the low pressure circuit takes less than 10 seconds for each of the two bleed nipples on the rear calipers. Tighten the bleed fittings to 8-10 ft.-lbs. using a six point 7/16-in. deep well socket.

If you have a relatively large air compressor, you can use it to power a Mityvac® MV6830 Vacuum Bleeder. (See fig. 2-4-7.) This solution is less awkward than the Motive Products unit because it attaches to the standard bleeder screws on the rear calipers rather than the non-standard hydraulic reservoir.

The bleeder is attached to each of the two low pressure bleed fittings on the rear calipers in turn. Apply vacuum to the fitting and then open the bleeder until fluid runs clear and air free. Close the fitting, transfer the bleeder to the fitting on the other rear wheel and repeat. See fig. 2-4-5 to identify the correct location of the low

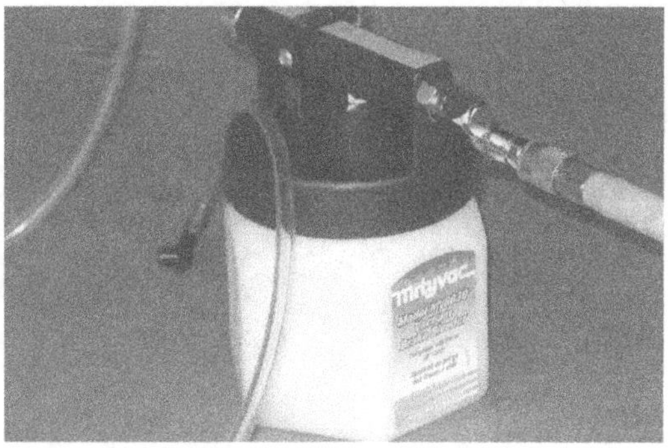

Fig. 2-4-7 – Mityvac MV6830 Vacuum Bleeder.

pressure bleed fittings. Make sure that the fluid level in reservoir #2 remains above minimum throughout this procedure.

Section 2-5
Change Trunnion Joint Oil

The trunnion joints are on the inboard ends of the rear halfshafts adjacent to the differential. When and if they leak, they do so into the differential leaving no outward indication of trouble. It is important to check them on an annual basis.

In order to do this procedure, you will need to position each trunnion joint in various positions to first drain and then fill it. You can do this by moving the car back and forth. Alternatively, chock the front wheels so that the car cannot move. Release the parking brake and place the transmission selector in neutral, then turn the ignition key to the run position so that the transmission actuator puts the transmission in neutral, then turn the key back to off.

Once you have the trunnion in the desired position, reduce the risk of injury or death while working underneath the vehicle. Disable the transmission by removing the gear shift cut out from the fuse board.

Working on one wheel/trunnion joint at a time, jack one of the trailing arms near the wheel end until the wheel is free to spin. Since the transmission is in neutral, you can now rotate the wheel (trunnion and driveshaft) to any desired position. I haven't tried to do this procedure without a lift, but it is possible in a pinch. Be careful!

Note: while raising the rear wheel and trailing arm with the bottle jack, the bottom surface of the trailing arm will change orientation as it moves closer to the chassis. This may cause the bottle jack to lose contact with the trailing arm, which will allow the wheel, trailing arm, and the rest of the car to come down with potentially lethal results. If in doubt, don't do it!

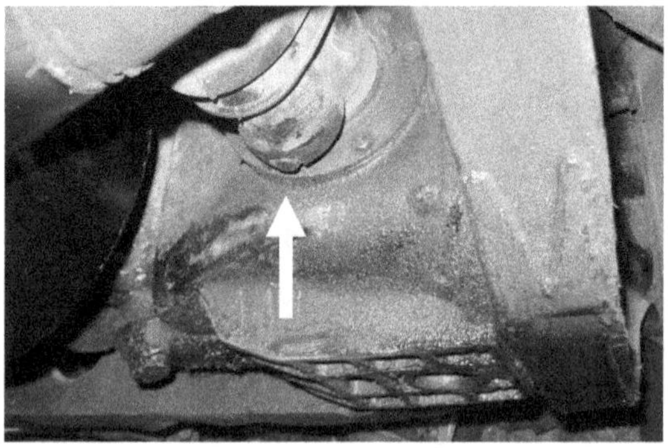

Fig. 2-5-1 – Drain plug location on left trunnion joint.

To drain a trunnion joint, position the wheel so that the filler screw is pointed down. Use a 1/2-in. A/F socket or box end wrench to remove this screw. Measuring the amount of lubricant coming out is a good idea as it will give you an indication of how fast they are leaking.

Reposition the wheel so that the filler screw is at approximately 10 o'clock. Using a large syringe, add 150 ml (5 oz.) of the same gear lube used in the differential. (I use Mobil 1 75W-90 Synthetic Gear Lubricant in a 50cc regular taper tip plastic syringe. The syringes are available from McMaster-Carr.) Replace the screw, then repeat the procedure for the other side and rest easy with well lubricated trunnion joints!

Section 2-6
Ignition Tune-up

On the whole, the Silver Shadow ignition system is very similar to other cars of the same period. Early cars prior to chassis no. 8742 use a twin breaker points setup, subsequent cars built prior to 1975 use a single breaker point set up and all remaining cars through the end of the Shadow II series use the Opus electronic system. Virtually every model year, the specifications for spark plug gap, dwell, spark timing and idle speed changed as well. (See Section 10-3 for year by year tune-up data.)

A full ignition tune-up consists of the following: replace spark plugs and ignition wires; replace distributor cap, rotor, points (non-Opus) and condenser (non-Opus); adjust dwell, idle speed, and ignition timing.

In practice, the majority of these components truly need service only every 15,000 miles, although it certainly does not hurt to check them more frequently. However, one thing to check on pre-Opus distributor equipped cars is the dwell.

Dwell is the number of degrees the distributor cam rotates while the points are closed. It is inversely proportional to the breaker point gap: decreasing the point gap will increase the dwell. If the dwell is too small, the magnetic field in the ignition coil will not have time to regenerate between firings and the spark will be weak. If the dwell is too large, the breaker points will not open consistently and misfiring will result. In addition, large dwell settings can overheat the coil. A dwell meter is the best method of adjusting breaker points.

On the single point distributor cars in particular, the dwell tends to drift out of spec on the low side in as little as 3,000 miles. This is a natural consequence of wear on the points creating a larger gap over time. Given the shallow profile of the eight sided distributor cam, there is little difference between a proper points gap and a gap that is too large to close reliably and consistently 160 times per second at highway speeds. If your engine begins to cut out, suspect incorrect dwell first!

If your dwell goes out in less than 500 miles and the points are pitted, it is time for a new condenser.

Replace Spark Plugs and Ignition Wires

The spark plugs can be changed with a 3/8-in. drive ratchet, a 3/8-in. universal joint, a 13/16-in. 3/8-in. drive spark plug socket and a variety of 3/8-in. extensions. The early and late cars are generally easier to access than the early to mid-Seventies examples.

For these cars, it may be to your advantage to remove the windshield washer bottle. Simply disconnect the two electrical and one liquid connection at the top and lift the bottle out completely.

Adjust the spark plug gap to the specification in Section 10-3 and coat the threads with anti-seize paste prior to installation. Use a torque wrench, if possible, and tighten to 15-18 ft.-lbs.

Replacing the spark plug wires requires an altogether higher degree of effort because of the way the wires need to be routed for a proper job. For the emissions equipped Shadows, so much hardware needs to be removed that it makes sense to change the valve cover gaskets at the same time. The service interval on both items seems to work out to about 9-15 years. This is covered in Section 3-1.

If your ignition wires are cracked and brittle, or if you can see stray high voltage coming from the insulation with the engine running in a darkened garage, particularly near the distributor, then it is time to replace them. Litigious types please note that this test should be performed at nighttime *with the garage door open.*

Replace Distributor Cap, Rotor, Points and Condenser

Replacement of the distributor cap and rotor is straightforward and does not need explanation. Fig. 2-6-1 can be used as reference for the correct spark plug wire sequence on the distributor cap.

In the case of both twin and single breaker point ignition systems, removal and replacement of the points and condenser is a simple matter of removing the nut(s) and screw(s) that attach the breaker points and condenser to the base plate. (See fig. 2-6-2.) Be extraordinarily careful not to drop any of these items into the lower workings of the distributor!

Reassembly is straightforward excepting one minor but important detail on the single breaker point system. The mounting post for the moveable portion of the breaker is sandwiched between two insulating washers. These washers have a flat side and a domed side. The domed sides are designed to face the breaker and ensure that the breaker arm is kept from touching the mounting post. (See fig. 2-6-3.) If the breaker arm is not reassembled with

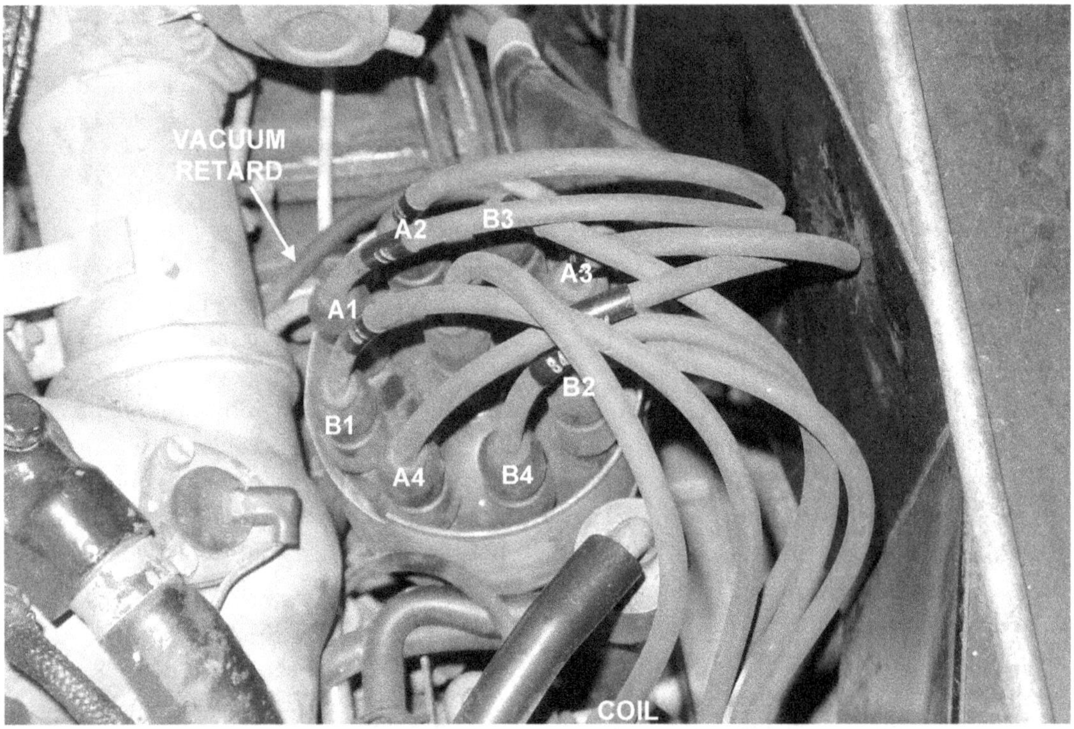

Figure 2-6-1 – Proper location of each spark plug wire on the distributor cap.

care, it will touch the post and ground the coil such that the car will not start.

If the car does not start after you replace the points, this is the first thing for you to check.

<u>*Adjust Dwell, Ignition Timing and Idle Speed*</u>

Connect an engine analyzer similar to the one shown in fig. 2-6-4. Since the battery is in the trunk, the red power lead on the analyzer can be connected to the alternator B+ terminal instead. The black ground lead can be connected to any grounded metal object in the engine compartment except the carburetors or fuel lines. The green lead is connected to the minus side of the ignition coil. (See fig. 2-6-5.) Ensure that all of these leads are routed away from the moving parts of the engine.

To reduce risk of injury or death while working with the engine running, chock the wheels and disable the transmission by removing the gear shift cut out from the fuse board.

Start the engine and bring it to operating temperature. Ensure that the air conditioning is turned off. The dwell will read the same regardless of engine speed, as long as the engine is running smoothly somewhere between the 500-1200 RPM range. Verify the RPM on the engine analyzer and then switch it over to dwell. Your desired dwell setting should be the high side of the specification shown in Section 10-3. The high side is chosen since the dwell drifts down as the points wear.

Adjust the distributor dwell adjustment screw to exceed the desired dwell setting and then reverse direction to arrive at the desired dwell setting. (See fig. 2-6-2 for location of the dwell adjustment screw.) There is quite a bit of hysteresis in this adjustment and approaching the desired value from the high side will all but eliminate the need to fine tune this adjustment after the car has been returned to service.

Ignition timing is dependent on RPM, so it is important to set the RPM to the speed

Figure 2-6-2 – Single point type distributor.

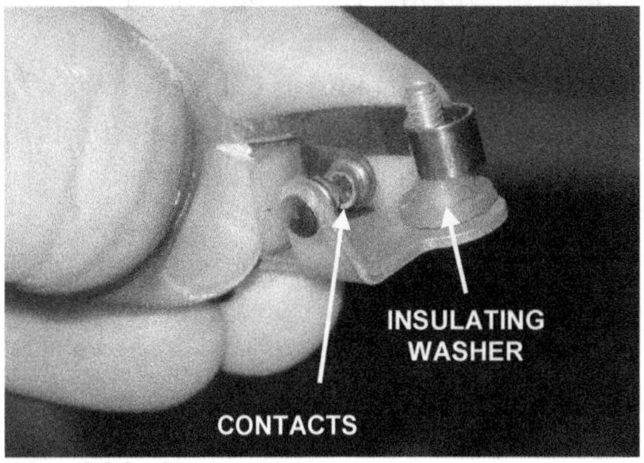

Figure 2-6-3 – Breaker points.

recommended in the "Ignition Timing" column in Section 10-3, not the "Idle Speed" column. First, it is important to disable the "vacuum retard" feature if it is present on the distributor. (See fig. 2-6-1 for the location of vacuum retard hose.) To do so, disconnect the vacuum hose at the distributor and plug the hose with a golf tee or a small Phillips screwdriver.

The idle speed adjustment is on the throttle linkage immediately to the right of the flexible air intake hose that connects to the choke housing. To adjust the idle speed, loosen the 2BA lock nut and turn the set screw while referring to the engine analyzer to achieve the desired RPM.

Attach an inductive stroboscopic timing light to the engine with the red and black leads connected to the same points used for the engine analyzer and the inductive pickup attached to the spark plug lead labeled "A1", which is the spark plug wire leading to the front cylinder on the right

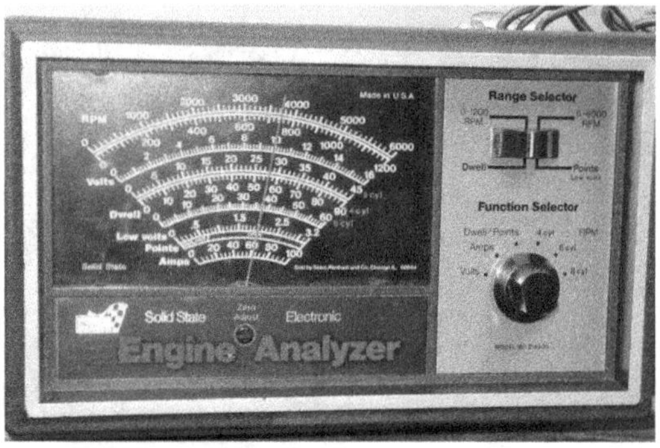

Fig. 2-6-4 – Engine analyzer set for dwell reading.

side of the engine. Some inductive pickups have an arrow on them that has to point toward the spark plug. If you have a very early car, ensure that the octane selector is in the fully advanced position before continuing.

Point the timing light at the timing pointer adjacent to the crankshaft pulley on the left side of the engine. If timing adjustment is needed, loosen the clamping screw on the distributor and rotate the distributor until the timing mark lines up with the timing pointer. *Be careful to keep your hands, sleeves, hair, tie, ascot, and equipment leads clear of rotating parts.*

For reference, the ignition timing is listed in Section 10-3, but only two hash marks will appear on the crankshaft: top dead center (TDC) and the correct number of degrees before top dead center (BTDC).* If you can see two marks on the pulley, but can no longer read the writing on them, the top mark as viewed from the left side is TDC and the bottom mark is the mark that you want to use to set the timing.

Tighten the distributor clamping screw and reattach the vacuum retard hose, if applicable. Adjust the idle speed to the proper value listed in Section 10-3 for "idle speed" and tighten the lock nut on the idle adjustment screw.

Disconnect all of the test apparatus and take her for a test drive. Check for smooth, abundant acceleration with no audible knocking or pinging from the engine.

* Note that the 1968 and 1969 model years have only one mark because ignition occurs at top dead center (TDC).

Section 2-7
Change Differential Oil

The differential on your Silver Shadow is a very robust piece of hardware. This does not mean you should neglect it. Even though the oil is not subjected to combustion gases, differentials do get hot and they also subject the oil to considerable pressure where the pinion gear meshes with the ring gear.

To reduce risk of injury or death while working underneath the vehicle, chock the wheels and disable the transmission by removing the gear shift cut out from the fuse board.

After a long run on the highway to warm the gear lube, place a drain pan under the differential. Use the small end of the hex adapter from the tool kit (same one used for draining the engine oil) in conjunction with a 13/16-in. socket and a 1/2-in. drive ratchet to remove the fill and drain plugs. Allow enough time for all of the old lubricant to drain out.

Replace the drain plug. Refill the differential with 5-1/2 pints or so of fresh gear lube. This is the approximate amount it will take before the oil begins to run out of the fill hole. I use Mobil 1 75W-90 Synthetic Gear Lubricant. Replace the fill plug.

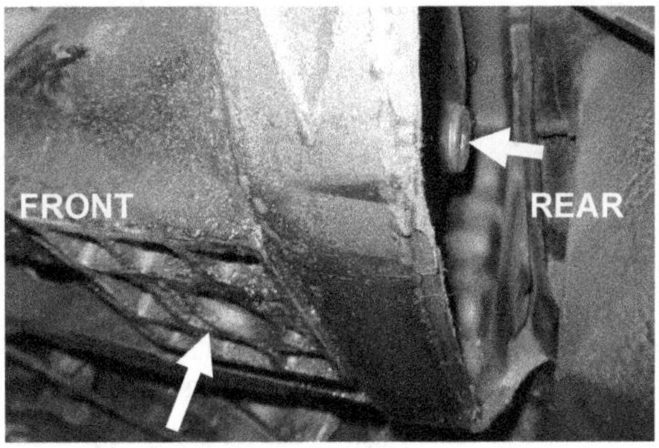

Fig. 2-7-1 – Drain and fill plug locations (l-r) on differential.

Section 2-8
Check Hydraulic System Function

Quite often referred to as "The Brake Pedal Test", this is one quick way to assess how well the accumulators are storing pressure.

1. Start the engine and run it until both brake pressure warning lights extinguish. (It should take less than 10 seconds.) Assuming that the warning lights are operating properly, if you fail at this step you have a major leak or the hydraulic pump is not pumping efficiently (or at all) on the indicated circuit.
2. Turn the engine off, turn the key to run, then depress (and release) the brake pedal 30-50+ times. The brake warning lights should not illuminate for at least 30 cycles. Assuming the warning lights are working correctly, if you fail at this step, the nitrogen is depleted in the accumulator sphere on the indicated circuit.
3. Restart the engine. The lights should extinguish in less than 10 but more than 1-2 seconds. If the light on the affected circuit goes out right away, this confirms that the nitrogen is depleted in that accumulator sphere and it is filled with brake fluid.

Figure 2-8-1 – Shadow I brake warning lamps.

Figure 2-8-2 – Shadow II brake warning lamps.

Any failures in any of the steps above require the attention of a professional and are beyond the scope of this book. Rebuilding accumulators is a normal maintenance item and not the end of the world. It just requires some specialized tools and should cost hundreds, not thousands, of dollars to remedy. Get it fixed now, no excuses—your safety depends on it.

Chapter 3
Engine

Section 3-1
Replace Valve Cover Gaskets and Ignition Wires

As your Shadow ages, two items tend to deteriorate at about the same pace: valve cover gaskets and ignition wires. Addressing either one of them requires a lengthy sequence of parts removal. Fortunately, the sequence is the same for both.

The valve cover gaskets are made from a cork composite and become compressed over time, leading to oil leaks. Aside from the untidy appearance, unaddressed valve cover leaks can compromise the integrity of the electrical connections on the starter culminating in a "failure to proceed."

Engine heat causes the insulation on the ignition wires to break down. The high voltage they carry begins to find ground paths other than the tip of the spark plug, resulting in weaker spark and engine misfiring, particularly in the presence of moisture. If your Shadow is not idling smoothly or you can see arcing around the

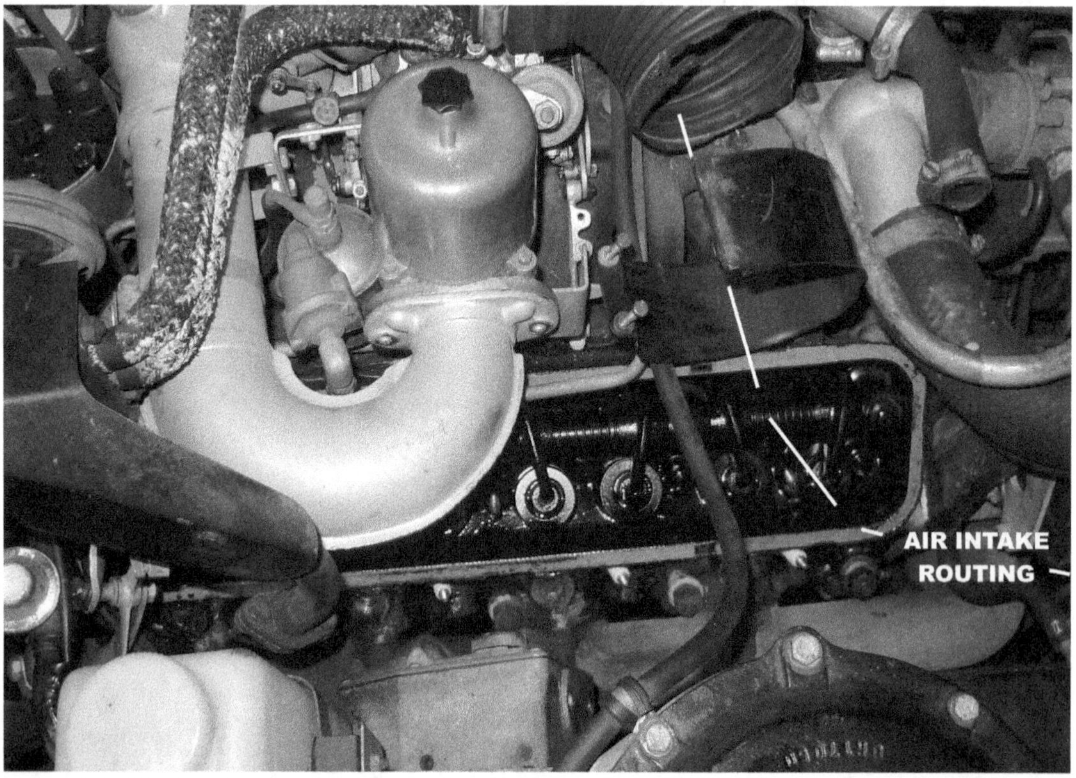

Figure 3-1-1 – "A" side rocker cover, "antlers" and majority of air intake hoses removed.

distributor with the engine running in a darkened (and well ventilated) garage, the time has come to change the ignition wires.

"A" Side

Begin with the "A" side, because it is easier to address than the "B" side. Remove the flexible air intake hose from the choke housing down to (and including) the air cleaner cover. The black air intake hoses can be freed by loosening the Jubilee clips that attach at the butterfly elbow and filter casting, and the machine screws that cinch the clip and bracket assembly in the center. This clip does not need to be removed from the lower intake manifold. Also remove the hot air intake duct from the fender apron by loosening one Jubilee clip and two wing nuts. Take great care during the removal of this part. You may find the rubber on most of the sections to be on the verge of disintegrating. A complete kit of replacement Crewe parts costs almost $600.

For later emissions-laden Shadows, you will need both a 3/4-in. stubby wrench and a wide variety of socket extensions for a 3/4-in. crows foot wrench to loosen the air manifold assembly, a.k.a. "antlers", from the exhaust manifold. Once you get in there, all four nuts do indeed loosen without too much protest. It helps to remove the windshield washer bottle for additional access. Once the exhaust end is loose, disconnect the other end at the flexible hose leading to the air pump and the antlers should come free.

To remove the "A" side ignition harness, start by disconnecting the ignition wires from the spark plugs and distributor. Remove the two acorn nuts on the valve cover and the bolt attaching the ignition wire boot bracket to the back of the head. The entire harness including the harness bracket can now be removed by pulling it forward out of the engine bay.

The valve cover is then easily removed. (See fig. 3-1-1.) You may find the old gasket compressed all the way down to the metal retaining lip on the inside of the cover. The gasket can be gingerly pried away. A gasket scraper used in conjunction with gasket remover spray will remove the adhesive residue. Replace the rubber bushings on the valve cover and the copper grounding clips on the valve cover gasket. Use Permatex No. 2 gasket sealant or similar on the cover and the facing side of the new gasket to hold it in place for reinstallation. (See fig. 3-1-2.) (Do not put sealant on the side of the gasket that faces the head. However, a thin coat of multipurpose grease on this side helps the valve cover assembly seal to the head.)

If you choose to fabricate a replacement ignition harness, consult Section 3-2 for specific recommendations.

If you are installing a replacement flexible air intake, note that the large hose kit includes two bands to protect it from abrasion. The long protective band slips over the air intake hose adjacent to the

Figure 3-1-2 – Cleaned rocker cover with new gasket and grounding clips in place.

bonnet hinge. The bracket that is attached to the lower intake manifold captures the short band.

The replacement hose kit does not include a resonator, nor will the old one fit in the new hose. With this significant restriction removed from the intake you may have to adjust the air-fuel mixture. (See Section 3-5.)

The "A" side takes about five hours to complete, which includes cleaning of all the parts. During reassembly, remember to use anti-seize paste on the "antler" connections and spark plugs in particular.

"B" Side

Take copious notes and photographs as you disassemble the "B" side to ensure that all the tubes and wires will return to their original position upon reassembly. To begin, the battery should be disconnected so that the alternator can be safely removed. On a few models, the purge line along with the "G2" filter should be removed to gain access. The "G2" filter is named after its replacement part number and is the smaller of two filters mounted on the left fender apron.

After disconnecting the alternator electrically, remove the bottom fastener which will release the belt tension. Remove the belt from the alternator pulley. While supporting the alternator with one hand, remove the long mounting bolt and spacer assembly at the top of the alternator, allowing the unit to come free. (See fig. 3-1-3.)

Remove the oil filler neck from the front of the engine. Next, remove the flame trap, which should be cleaned with kerosene and dried with compressed air in a well ventilated area. The flame trap is a small casting attached to the rear of the oil filler neck.

Now the "U" shaped manifold piece from the "B" bank carburetor must be

Figure 3-1-3 – Oil filler fasteners can be accessed now that the alternator has been removed.

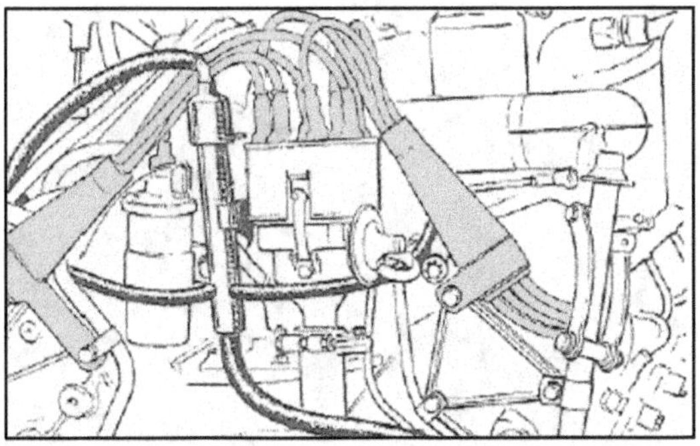
Figure 3-1-4 – Orientation of ignition wire boot brackets. (Illustration supplied courtesy of Bentley Motors, Inc.)

removed to allow removal of the PCV (positive crankcase ventilation) hose which is normally attached to the flame trap. For later Shadows, the "solenoid shelf" needs to be removed in order to gain access to the manifold piece in question. Make sure you record the exact layout of the vapor lines to the components on the solenoid shelf. These vary quite a bit from year to year.

The two "stove pipes" that direct heated air to the choke thermocoil mechanism are removable. The nuts at the manifold end tend to be corroded and a shot of PB Blaster a few days in advance will usually help. At last you can now remove the "antlers" and the wire harness. A note on stove pipes: since rusty ones will disintegrate no matter how carefully they are removed, you may want to inspect yours before beginning this project and order parts accordingly to avoid delaying the reassembly process.

For many Shadows, the "antlers" will remain trapped under the A/C and low pressure hydraulic lines, even with all of the connections loose. However, there is enough room to slide the "antlers" off the front mounting stud, which is enough to free the ignition wire conduit.

The bolt that attaches the ignition wire boot to the back of the head on this side is hard to see. (See fig. 3-1-4.) It is the larger and lower of the two and requires a 1/2-in. open end wrench. With this removed, it remains nearly impossible to remove the

Figure 3-1-5 – Heater tubes block access to "B" side rocker cover on later Series I Shadows

Figure 3-1-6 – Heater tubes and "B" side rocker cover removed. Note the use of paper towels to prevent debris or small parts from getting into sensitive areas.

ignition harness in one piece. Instead, ensure that the ignition wires are properly marked at the distributor end and record their location on the distributor cap. Remove the wires from the distributor cap and cut them as close to their connectors as possible, thus preserving the lengths that you will need to measure later. Remove and save the wire labels, if any. Then, carefully tug the wires through the boot, pulling only the wires and metal conduit forward out of the engine bay. (See Section 3-2 for additional detail. See fig. 3-2-2 for a photo of the metal conduit and wire assembly.)

The easiest way to remove the "B" side valve cover with minimal disturbance is to remove the two hard line heater tubes that carry coolant from the water pump to the heater core. (See fig. 3-1-5.) To do so, some of the coolant needs to be drained by opening the tap on the lower left side of the radiator. With these heater tubes removed, the valve cover is finally free. (See fig. 3-1-6.)

Reassembly is straightforward, with only a few fine points. Refurbish the valve cover as described in the "A" side with new gasket affixed with Permatex No. 2, new rubber bushings and copper grounding clips. For further detail on installing the ignition wires, refer to Section 3-2.

You may want to clean, prime and repaint the cast portions of the PCV hose and flame trap as well as the thermocoil housing and ignition conduits with high temperature black glossy paint. Replace the o-rings on the hard line heater tube connections at the water pump, fiber washer at the stove pipe thermocoil connection, and the gasket at the oil filler casting. If the carburetor float chamber connections are disturbed, replace the specially shaped fiber washers on these as well.

In order to reinstall the "solenoid shelf" you will likely need to use a magnetic insert inside a 7/16-in. socket to reinstall the retaining nuts. Remember to use anti-seize paste on all of the threaded connections.

The "B" side takes upwards of 32 hours to complete including cleaning and painting.

Figure 3-1-7 – List of Materials

Parts

2	Valve cover gaskets (UE4324R)
1	Oil filler "triangle" gasket (UE75131)
6	Valve cover grommets (UE42838)
8	Copper grounding clips (UE34258)
2	Heater hose to water pump o-rings (PS110173PA)
1	PCV fitting o-ring (PS110173PA)
1	PCV fitting o-ring (UE35022)
1	Flame trap o-ring (CK612)
1	Flame trap o-ring (CK540)
1	Stove pipe/thermocoil fiber washer (UE5994)
2	Float chamber fiber washer (RE7297)

Consumables

3 gal. Prestone antifreeze mixed 50:50 with distilled water
Permatex Gasket Remover
Permatex Anti-seize Paste
Permatex No. 2 Gasket Sealant
Parts degreaser or kerosene
Gloss black high temperature paint

Optional

1	Air intake hose, 3-pc. kit (RH2802)
1	Hot air intake (UE38027)
1	Stove pipe, manifold to butterfly housing (UE35025R)
1	Stove pipe, manifold to thermocoil (UE35026R)

Section 3-2
Fabricate and Install a Set of Ignition Wires

It is not possible to do a proper job of replacing ignition wires without removing the metal conduit that holds the wires neatly along the valve cover. If your Shadow has an air pump, this conduit cannot be removed until the air injection "antlers" are removed first. If you are going to do all this, follow the instructions in Section 3-1 for removing the air injection plumbing and while you are in there replace the valve cover gaskets too.

Once you have all of the air injection items removed (or if they were never there in the first place) removal of the ignition harness is simple:

1. Remove the acorn nuts that attach the conduits to their respective valve covers.
2. Remove the 1/2-in. A/F screws that attach the rubber ignition wire boots to the rear face of each cylinder

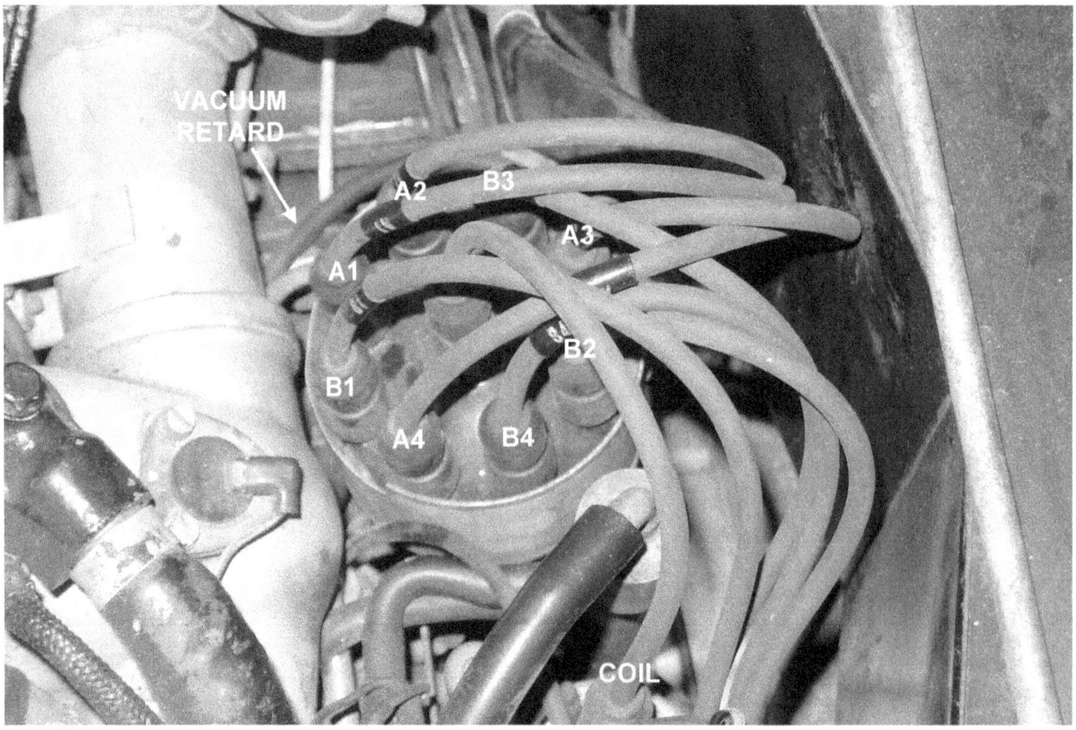

Figure 3-2-1 – Proper location of each spark plug wire on the distributor cap.

head. (See fig. 3-2-3.)

3. Ensure that the wires are properly marked at the distributor end and record their location on the distributor cap.
4. Disconnect the wires from the distributor cap and from the spark plugs. Lift each harness from the engine bay.
5. In order to remove the rubber boots and labels from the old wires, cut the distributor wires as close to their connectors as possible, thus preserving the length that you will need to measure later.

Crewe supplied replacement ignition harnesses vary in availability and are fairly pricey when they are in stock. It is possible to select very durable materials with OE appearance that provide proper resistance for good spark and reasonable points life in addition to offering better value. On this last point, the cost of materials used in this section is approximately half that of a commercial reproduction and one sixth of an OEM replacement set.

The solution consists of 7mm Delphi-Packard copper core wire with Beru terminations that thread onto the wire without need for a crimping tool. For later transistorized ignition (Opus) Shadows, you will need to use resistor wire rather than copper core. You can reuse the two large rubber boots at the rear of the heads and repaint the metal conduits. The Beru spark plug caps use the superior M4, rather than SAE type connections which means you have to remove the acorn nut from the spark plugs to expose the thread. These connectors make a "zip" noise when properly connected rather than the usual "click". Note that you cannot use Autolite plugs with this set up because the acorn nut connection is not removable. I have had good luck with NGK BP5ES spark plugs. (BPR5ES for later Opus-equipped Shadow and Shadow II.) If you find that the BP5ES/BPR5ES tend to foul with your particular driving patterns, use the hotter BP4ES/BPR4ES instead.

If the old ignition wire set has labels, they can be removed with a little coaxing if you have not already done so. The old wires will slip through the rubber boot and then can be eased out of the metal conduit. Measure each individually, adding a safety factor of two inches. As a sanity check, each wire should be about 4 inches longer or shorter than its neighbor, allowing for minor variation due to its location on the distributor. *For the sake of your mental health, you definitely want to get this right the first time!* With a modest safety factor and careful measurement, you will use almost exactly 25 ft. of wire to fabricate all eight spark plug wires and the coil wire.

Twist the spark plug caps onto the wires and insert them into the metal conduit leaving 6-1/4 inches from the tip of each spark plug cap to where it enters the

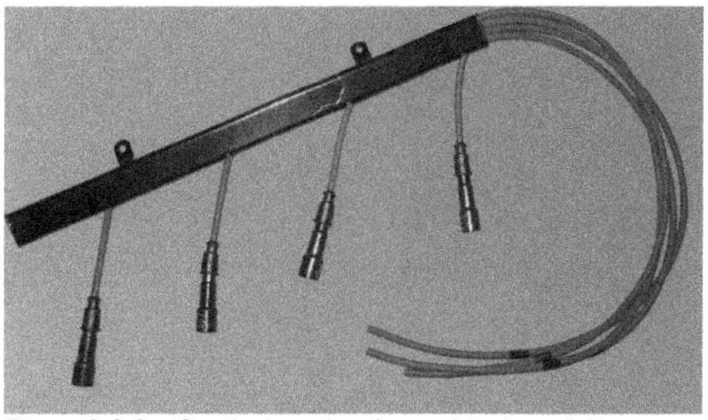

Figure 3-2-2 – Spark plug wires in repainted conduit.

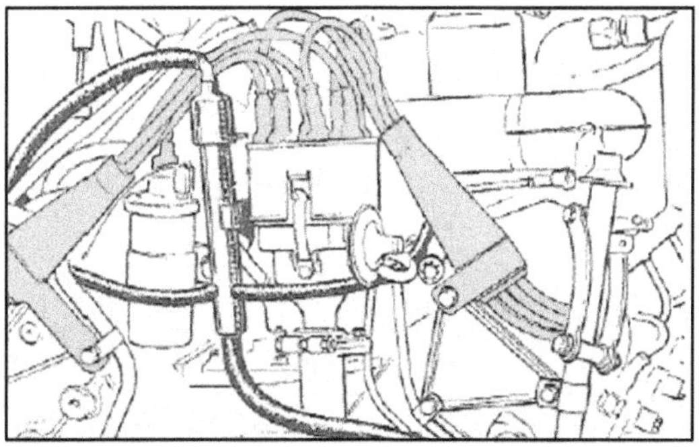

Figure 3-2-3 – Orientation of ignition harness brackets. (Illustration supplied courtesy of Bentley Motors, Inc.)

conduit. This allows precisely enough wire to have uniform wire loops that duplicate factory appearance. (See fig. 3-2-2.)

Check each wire for continuity from the distributor end to the spark plug connector. This rules out any electrical faults that will be hard to reach later.

Install each subassembly to its respective the valve cover, snaking the temporarily labeled wires toward the distributor. Slide each quartet of wires into the large boots, removing and replacing the labels as you do this so that the labels end up near the distributor.

With the boots in approximate installation position, reattach them to their mounting positions with the 1/2-in. A/F screws. This is a bit finicky for the "B" side and might require a small pair of hands. (See fig. 3-2-3.)

Finally, trim the wires to length where they meet their assigned terminals at the distributor and screw on the connectors.

Section 3-3
Diagnose Faulty Anti 'Run-on' Solenoid

This section is 90% story and 10% fix relating to an odd but not uncommon failure you might experience on 1973 and newer Shadows.

On a cold, sunny November day in Detroit, I decided to take my Shadow out for some exercise. She ran beautifully, benefiting from a recent major tune-up. As I emerged from my neighborhood, I dipped into the throttle to clear the cobwebs from the engine. I could see a wisp of hydrocarbons in the rearview mirror as she gathered speed. I ducked back into my neighborhood, preparing to turn onto my street, realizing abruptly with an armful of heavy steering that the engine had quit dead. After a few moments in awkward silence, I got her restarted for a short time, managing to achieve sufficient velocity to coast (through a stop sign) back to my driveway.

Because of the recent tune-up and the resolute manner in which the engine had died, I assumed ignition to be the root cause of the "failure to proceed". Once back in the unheated garage, the engine would start easily from cold and run smoothly for four or five minutes, then stop dead without sputtering or any other form of protestation. It would turn over but fail to restart until the engine rested for a few hours.

I spent a week's worth of spare time checking the system in order of perceived likeliness: points, coil, ignition wires, ballast resistor, low tension leads, distributor cap, rotor, etc. Finally, I instrumented both the high and low tension sides, started the engine and waited for it to die so that I could observe all of the vital information as it passed from 600 RPM idle to dead stop. As the engine slowed, the dwell remained completely stable, there was sufficient coil energy to trip the timing light, and there was no sudden change to the voltage on the low tension side of the coil as it had already stabilized to 7 Volts in the minutes prior. Clearly the problem wasn't ignition related. It was time to move on.

I checked the operation of the fuel pump. Disconnected on the outlet side, the pump seemed to move fuel with enough force and volume to sustain an idling engine indefinitely. There were further tests I could perform, but this avenue of pursuit appeared unlikely to reveal the problem. It seemed to me that weak or intermittent fuel delivery should cause sputtering or at least a higher speed lean idle as the float chambers empty.

Another week went by. I studied the layout of the various weakener system solenoids to see if I could contrive a situation in which the vehicle would run like a watch for four or five minutes and then stop. The weakener system is designed to create a partial vacuum on the float chambers at part throttle cruising speeds, which reduces the differential pressure from the float chambers to the intake manifold, thus reducing the fuel flow into the engine. Perhaps something was creating a major vacuum at all speeds instead, thus cutting off fuel completely?

As luck would have it, there is a solenoid placed in the weakener circuit to do just that, but only when the ignition is switched off. It is called the anti 'run-on' solenoid and is designed to prevent the engine from dieseling by cutting the fuel supply after the ignition is switched off. (See fig. 3-3-1.) This vacuum circuit diagram is representative for 1973 emissions equipped Shadows and derivative models. Model year differences are discussed below.

To prevent driveability concerns in cold weather, the weakener cut-off solenoid valve is designed to circumvent the weakener circuit and provide full atmospheric pressure to the float chambers (and thus a richer air-fuel ratio at cruising speeds) until the weakener cut-off valve switch (a.k.a. bi-metal switch) reaches a temperature above 54-61°F.

If the anti 'run-on' solenoid failed in the closed position and the intake manifold was sufficiently cold to switch open the weakener cut-off solenoid valve, perhaps the engine would run normally until the weakener valve cut-off switch warmed and closed the valve? At that point, a vacuum would form rapidly in the float chambers and the fuel supply to the engine would be cut off.

According to the Workshop Manual, I first ensured that there was battery voltage present at the anti 'run-on' solenoid when the ignition key was on. Next, I detached the solenoid end of the hose from the anti 'run-on' solenoid to the solenoid end of the 'Tee' piece and connected a longer hose to the solenoid. With the ignition switched on, it should have been possible to blow into the solenoid. It was not! (See fig. 3-3-2.)

As a further test, I pulled the wire from the bi-metal switch on the choke housing while the engine was running. This killed the engine instantly. After the engine had warmed up and died of its own accord, I simulated a cold engine by shorting the bi-

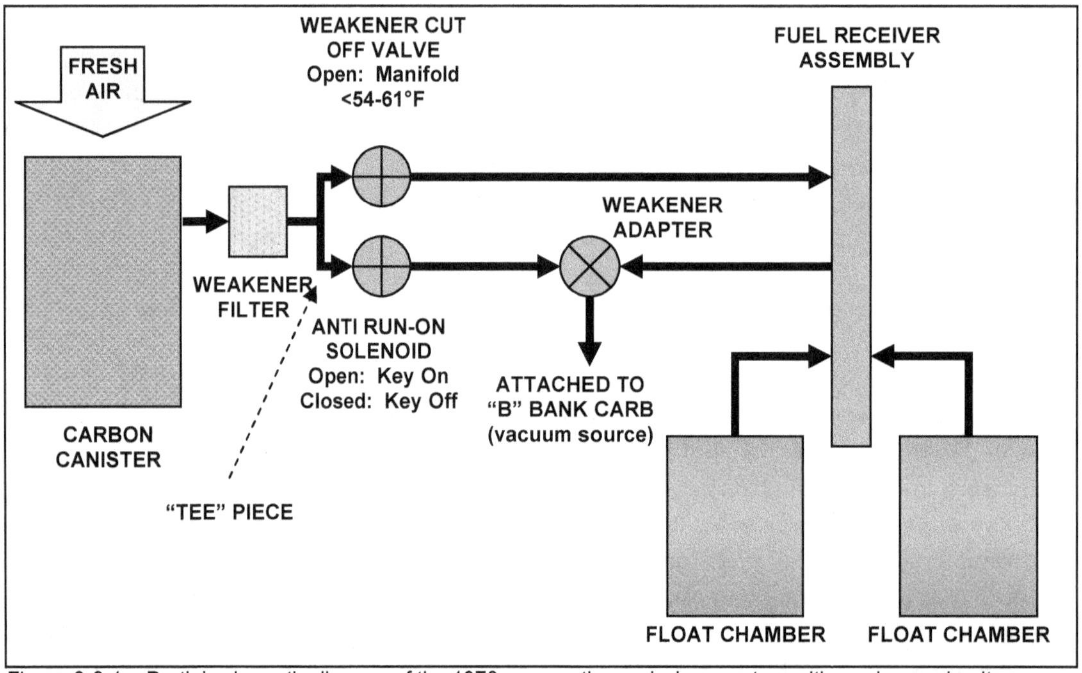

Figure 3-3-1 – Partial schematic diagram of the 1973 evaporative emissions system with weakener circuit.

metal switch to ground using a small jumper and was able to restart the engine immediately. (See fig. 3-3-3.) This series of four tests proved conclusively that the anti 'run-on' solenoid had indeed failed. Note that if the ambient temperature is above 54-61˚F, a Shadow with this condition will not start at all until the jumper is attached or, alternatively, the acorn shaped float chamber pressure tapping nut on the "A" side is loosened. (See fig. 3-3-4.)

Figure 3-3-2 – Anti 'run-on' solenoid location.

Fortunately, the anti 'run-on' solenoid is on top of the engine on the "B" side. To remove it, disconnect the two electrical leads, remove the two mounting screws and pull the two rubber hoses from either side. Installation of the replacement part (UE38169) is simply the reverse.

You may need to replace the two hoses as they become brittle with age. Also, the replacement part had somewhat larger diameter tubes that required a bit more stretching of the tube material. Suitable hose material is available at local auto parts chains and can be cut to length as you reassemble the system in the engine compartment. (Or if cost is no object, replace the hose material with Viton® tubing. See Section 3-4.)

As with most things emissions related, there is quite a bit of year-to-year variation. Prior to 1973 and on any fuel injected car, the solenoid was not used. From 1974 to 1980, the configuration changed so that the

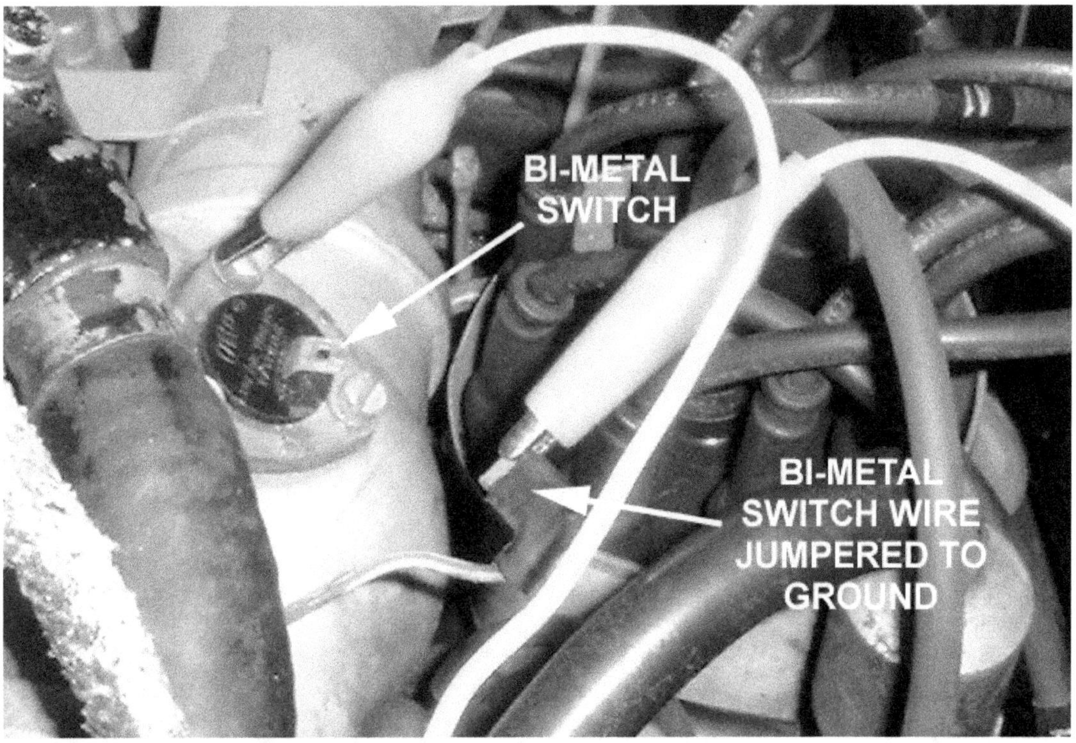

Figure 3-3-3 – Bi-metal switch behind choke housing is disconnected and the wire is grounded to simulate a temperature of less than 60˚F.

anti 'run-on' solenoid is connected to a vacuum tap on the manifold. In these applications, the solenoid *opens* with ignition off and *closes* with ignition on. (This is opposite to 1973 *and a different part number!*) Given the location of restrictors in the various lines, it is not clear if a vehicle with a solenoid failure in the open position would run well (or at all) with the weakener cut-off jumper test shown in fig. 3-3-3. An alternative get home strategy would be to disconnect (and leave open) the vacuum line that runs from the anti 'run-on' solenoid to the 'Tee' piece.

With this experience added to your mental portfolio of possible roadside fixes, consider adding a circuit tester and a small jumper lead to your emergency kit.

Figure 3-3-4 – Acorn shaped pressure tapping nut on "A" side carburetor float chamber.

Section 3-4
Replace Vacuum Lines

When working in a Shadow engine compartment, you will inevitably disturb an old, brittle vacuum line which will necessitate replacement.

My criteria for tubing material are for it to have the longest life possible while keeping an OEM appearance. Viton® tubing is a great solution. It is a proprietary material used in many critical automotive applications including high pressure diesel engine fuel systems. It is highly resistant to heat and all manner of volatile chemicals and will likely outlive the rest of the car. However, it is very expensive -- the tubing required to redo a typical Shadow engine bay minus the large and long evaporative canister lines is $100.

The next best material is silicone, which is less than one-fifth the cost. In either case, purchase the material with a hardness of 70A-75A so that the walls are stiff enough to resist collapsing under vacuum. Both of these types of tubing are available from industrial supply companies such as McMaster-Carr.

Stock five feet of each size and replace aging vacuum hoses as needed in the course of doing other maintenance. (See fig. 3-4-1.) If you choose to replace all of the hoses at once, take extreme care to remember where everything goes. Incorrect connections can cause a wide variety of issues: hot or cold idle stability, driveability at cruising speeds, lean stumble or misfire, poor fuel economy, idle flaring, and/or backfire on deceleration, to name a few. It is best to cause one problem at a time and correct it rather than causing several simultaneous problems with very similar symptoms that will even confound the experts!

Replacement Hoses

1/8-in. I.D., 1/4-in. O.D.

- Vacuum retard diaphragm to tee-piece
- Vacuum retard switch to tee-piece
- Vacuum retard switch to adapter

3/16-in. I.D., 5/16-in. O.D.

- Tee-piece to hot idle mechanism
- Adapter to vacuum tree
- Vacuum tree to vacuum tap
- Purge restrictor to gulp valve tap

1/4-in. I.D., 1/2-in. O.D.

- Anti 'Run-on' solenoid to weakener body
- Anti 'run-on' solenoid to tee-piece
- Cruise control

Figure 3-4-1 – Vacuum hose diameters.

Section 3-5
Carburetor Tuning

Many techniques have been devised for fine tuning carburetor balance in V8 powered Rolls-Royce and Bentley motorcars. Cal West's original article on the subject published in the 87-2 issue of the *Flying Lady* is certainly a good one. The problem is obtaining a Carbon Monoxide (CO) meter to do an accurate job. This section details the use of a low tech alternative, known as the Gunson Colortune. This device is a spark plug with a transparent base that can be installed temporarily in order to view the actual color of the flame in the combustion chamber and from that infer the air-fuel ratio.

All of the same principles described in Cal's article apply here. If your Shadow is running like it is out of tune, you need to verify that *all* of these things are 100% perfect before you touch the carburetors: ignition wiring, coil, points, condenser, ignition dwell and timing, spark plug condition and gap, choke and stove pipes, fuel pump, fuel filters, vacuum and evaporative emission hoses*, hot idle compensator, exhaust gas recirculation*, air filter, air pump* and vacuum retard tap*. The only item even less likely than the carbs to be out of whack is the weakener* adjustment, which is covered in Section 3-6.

---

* Items with an asterisk were introduced as running changes over the life of carbureted V8 and may not apply to yours depending on the model year and country of delivery.

The following also assumes that the throttle linkage has not been disturbed and the throttle plates close synchronously. To tune the carburetors, pick a 60-80°F day and perform the following operations in sequence:

1. If the engine is not already warm from the above checks, run the engine until the thermostat has been open for five minutes. You can tell the thermostat has opened by a sudden shift in the temperature of the upper radiator hose from comfortable to too hot to touch.
2. While waiting for the engine to warm up, ensure that the A/C is switched off and that the choke has opened all the way. Adjust the idle speed to the correct value specified in Section 10-3, if necessary.
3. Stop the engine. Loosen the locking nut on the each of the air-fuel mixture set screws. Fig. 3-5-1 illustrates the "A" side mixture screw and is labeled #20 "Jet adjusting screw". "B" side is similar. I tape a 6BA wrench to the end of a screwdriver to reach the locking nut; Rolls-Royce specifies the special tool RH8050.
4. Using a long screwdriver, turn each mixture screw (one per carburetor) counter-clockwise until it is just loose against the lever arrangement that adjusts the height of the jet.

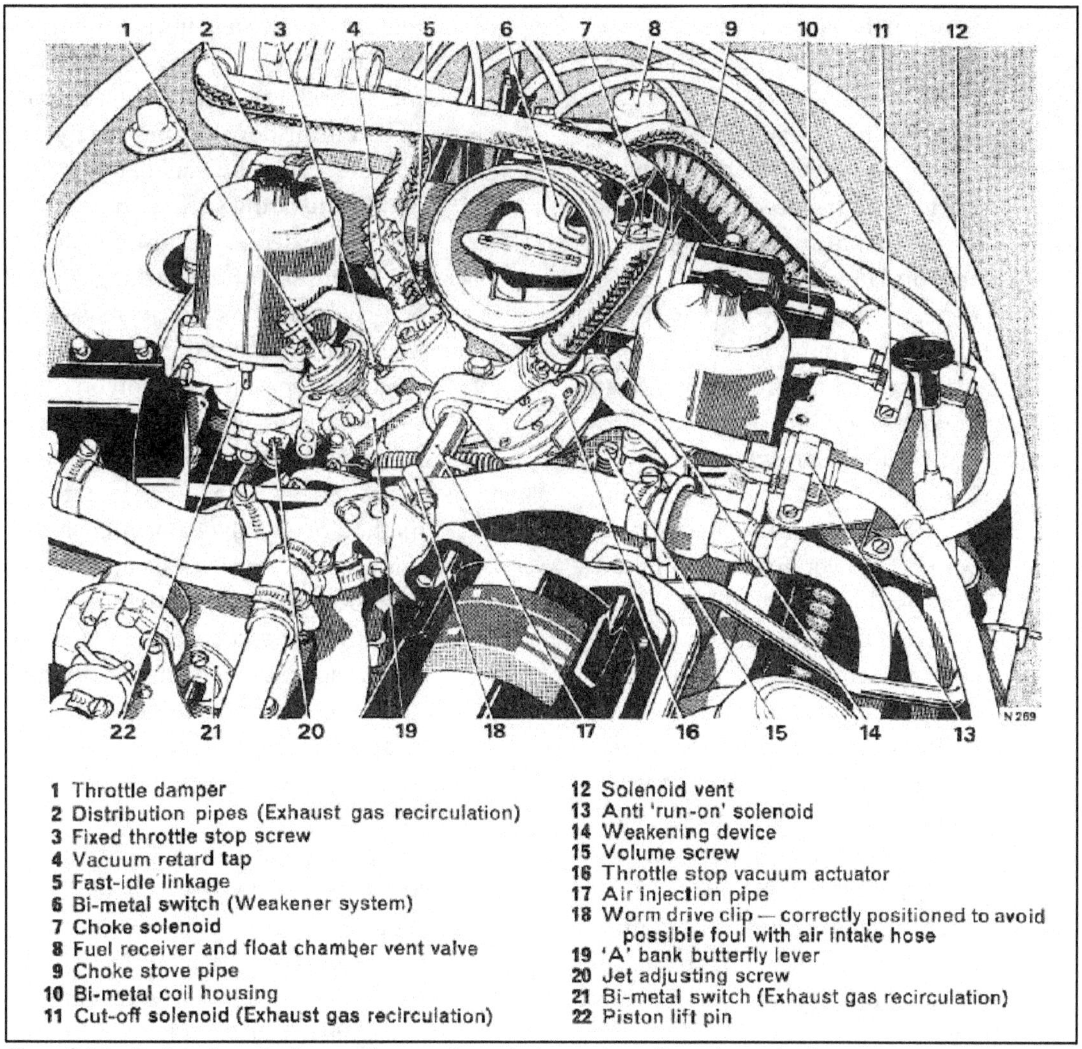

Figure 3-5-1 – Partial schematic diagram of evaporative emissions system with weakener circuit.
(Illustration supplied courtesy of Bentley Motors, Inc.)

Make sure that you do not loosen it so far that the screw falls out.

5. Turn each mixture screw clockwise 2-1/2 turns.
6. Turn each volume screw clockwise all of the way and then turn counter-clockwise 1-1/2 turns. Fig. 3-5-1 illustrates the "B" side volume screw (item 15). The "A" side is similar.
7. Starting with one carburetor, unscrew the black plastic damper cap and plunger and place on a clean work surface. It is important to establish two separate areas for each of the two sets of carburetor parts. You cannot mix and match them without causing headaches.
8. Remove the suction chamber by removing the four 2BA screws at the base. You may need to disconnect some of the weakener plumbing to access the screws on the "B" side carburetor.

9. Remove the large spring. It is very soft, so be careful not to stretch it out of shape.
10. Remove the air valve piston and needle assembly. Be very careful not to bend the needle. Dump the damper oil in an ecologically appropriate place.
11. Repeat disassembly steps 7-10 for the other carburetor.
12. Adjust each mixture screw so that the corresponding carburetor jet is flush with the bridge piece. (See fig. 3-5-2.)
13. If the carburetors and linkage are perfect, and you followed all the above steps, you should have balanced the carburetor volumes and air-fuel mixtures at idle. In reality, this merely gets you close.
14. Reassemble each carburetor with the exception of the damper cap and plunger ensuring that all parts are extremely clean. Use kerosene and a lint-free cloth, if necessary.
15. Use the same engine oil used in your crankcase to refill the dampers. Fill to approximately ½-in. from the top.
16. *To reduce risk of injury or death while working underneath the vehicle, chock the wheels and disable the transmission by removing the gear shift cut out from the fuse board.* Start the engine.
17. With the damper plungers still removed, place a length of tubing to your ear and to the opening where the damper cap screws in. You can fine tune air flow balance via the volume screws by equalizing the amplitude of the sucking noise. Clockwise decreases air flow (and the amplitude of the sucking sound) and counter-clockwise increases it. Make sure to place the end of the tubing at the same position on each suction chamber opening in order to minimize measurement error. Refer to Section 3-7 for an even better method of air flow balancing. Note: this adjustment is just as important to smooth idle as the mixture adjustment described below because it sets the volume of air-fuel mixture going into the four cylinders that this carburetors feeds.
18. Stop the engine. Replace each damper plunger to its correct carburetor and thread in the caps.
19. Remove the A4 spark plug and replace it with the Gunson Colortune plug. Remember to only finger tighten it; otherwise you may crack the transparent base. Because it is only finger tightened, you should only use this device in a no-load condition. (i.e., you can rev the engine in park, but if you drive around with it in there, it will likely pop out and damage the threads in

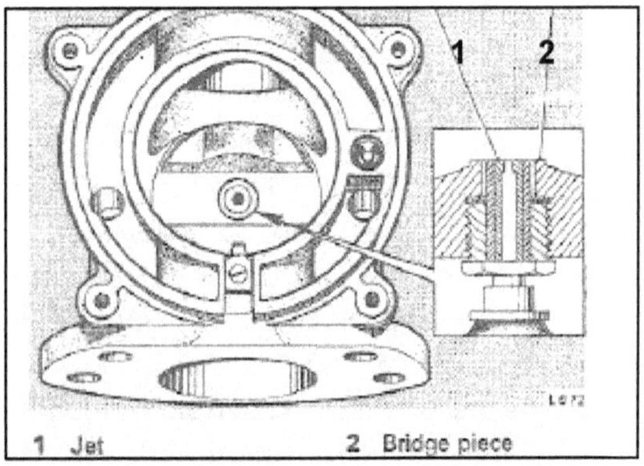

Figure 3-5-2 – Initial jet height is flush with bridge piece. (Illustration supplied courtesy of Bentley Motors, Inc.)

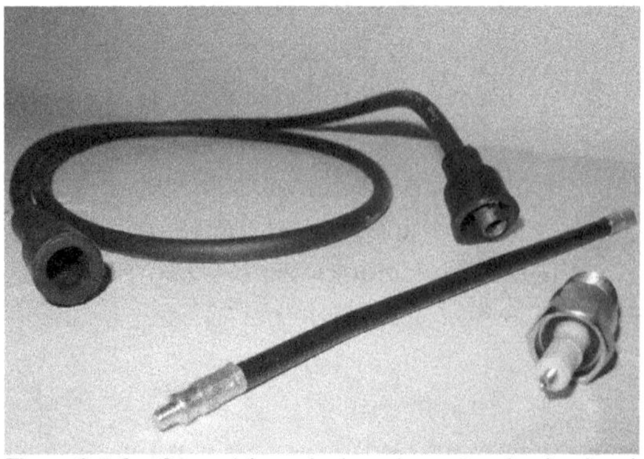
Figure 3-5-3 – Spare A4 spark plug wire, extension lead and Gunson Colortune.

the cylinder head.)

20. The Gunson Colortune comes with an extension lead that threads into the plug. Because of the peculiarities of OEM Rolls-Royce spark plug wire arrangements, it is easier to connect to this extension lead using a spare ignition wire attached to the A4 post on the distributor. (See fig. 3-5-3.)
21. Note: since the "A" side carburetor feeds cylinders A1, B2, B3 and A4, you can choose any of these cylinders to view the combustion color. For most model years, however, A4 makes the most sense in terms of accessibility and visibility.
22. Start the engine. Observe the combustion color. It should be Bunsen burner blue and flashing steadily (not misfiring) when the engine is idling. Grab the throttle linkage and gradually raise the engine speed to 2000 RPM or so. As the engine speed rises, the color should change to a bright orange rich condition, which means the damper is functioning properly and creating a temporary rich condition. Once the engine has stabilized at a steady 2000 RPM, the color should return to Bunsen burner blue, which indicates that the carburetor piston has lifted as it should. Release the throttle linkage and allow the engine to return to idle.
23. To adjust the mixture, turn the mixture screw clockwise (richen) until the combustion color *just* turns orange. *Then, while counting the number of turns it takes,* turn the mixture screw counter-clockwise until the Bunsen burner blue flashes just start to become irregular. (This is the onset of lean misfire.) Now, turn the mixture screw clockwise *half* that number of turns.
24. Replace the Gunson Colortune with the original spark plug and torque it to 15-18 ft-lbs.
25. Repeat steps 19-24 for the "B" side carburetor. B1 is likely the most convenient, but for reference, cylinders B1, A2, A3 and B4 are all fed by the "B" side carburetor.
26. Tighten the mixture screw locking nuts on both carburetors.
27. Re-adjust the idle speed to the value specified in Section 10-3, if necessary.

The Gunson Colortune is available in the U.S. for around $50 from a variety of sources.

Section 3-6
Weakener Gasket and Adjustment

This is another issue you may experience that is 90% story and 10% fix. After performing the valve cover gasket replacement procedure outlined in Section 3-1, my Shadow seemed to be losing power at steady cruising speeds. It felt like a lean mixture hesitation and I suspected it had something to do with excessive depression in the float chambers. To verify my suspicion, I built a 16-in. water manometer for $8. To build your own, refer to Section 9-4.

With the manometer connected to the pressure tap on the "A" side float chamber (Fig. 3-6-1), I found that the float chamber depression at 2,500 RPM was over 5 inches of water and that the power loss occurred any time the depression exceeded 4 inches. (Approx. 2-in. of H_2O is normal. Consult Section 10-3 for model year specific data.)

I checked all of the evaporative emissions plumbing that the Workshop Manual recommends in cases like this and after finding nothing, I jumped to the erroneous conclusion that the 3/32-in. Allen screw adjustment on the side of the weakener was out of whack. Such as it was, turning the screw actually had no effect on changing the float chamber depression.

After pulling what remains of my hair out, I found the cause to be the weakener gasket leaking from the EGR chamber to the weakener chamber. This gasket was a deteriorated item that I had replaced previously, but apparently it had not sealed.

Weakener implementation varies over the years. My 1973 Shadow has a weakener housing that also has a separate chamber to generate an EGR vacuum signal from an adjacent set of drillings in the carburetor body. Since the strength of the depression required to activate the EGR is fifteen times that needed to create a proper depression in the float chambers, a gasket leaking between the two areas would cause an outrageously high float chamber depression at part throttle regardless of how the weakener was adjusted. (See fig. 3-6-2.)

I found the mating surface of the

Figure 3-6-1 – Acorn shaped nut covers pressure tap on "A" side carburetor float chamber.

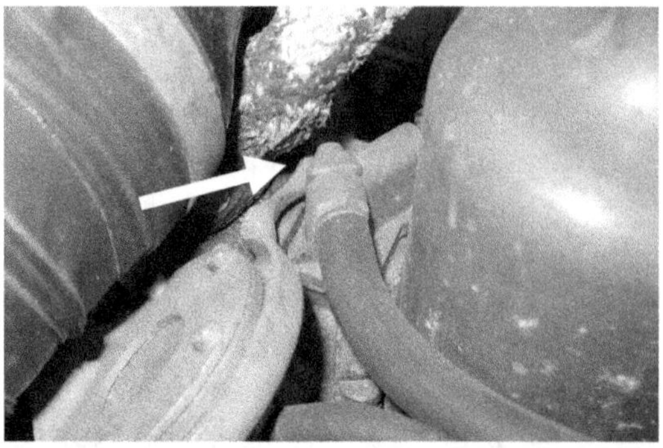

Figure 3-6-2 – Weakener housing on the "B" side carburetor adjacent to the intake hose.

weakener housing to be cupped by approximately 20 thousandths of an inch. Using a 9-in. "flat mill bastard" file, I carefully removed material from the mating face by moving the part across the file while ensuring that the entire face was in contact at all times. The moment the entire face was shiny, filing was complete. I also coated both sides of the new weakener gasket with high temperature RTV sealant and replaced.

The moral of the story: *do not* adjust your weakener; they do not magically misadjust themselves. If you do anyway, lock the adjuster threads using Loctite or Permatex green thread locking compound when you have completed the adjustment.

Section 3-7
Better Carburetor Balance

On most British sports cars equipped with multiple SU carburetors, a Unisyn flow meter is used to compare the idle air flow into each carburetor, which can then be adjusted accordingly. Unfortunately, the tool is designed to butt against each carburetor individually. Given the layout of the Silver Shadow intake manifold this is not practical (or in the case of the 4-bbl downdraft Camargue, not necessary).

If you stop and think about it, the SU carburetor itself is a precision measuring device, moving the variable venturi up and down in direct proportion to the air flowing through it. Measuring the height of each piston simultaneously provides you with a reference to balance the idle air flow.

Accurate measurement of the carburetor piston heights is achieved with a pair of 1-in. range dial indicators (with 0.001-in. gradations) securely mounted to each carburetor suction chamber with a threaded adapter collar. (See fig. 3-7-1.) These adapter collars can be ordered through Sherbourne Mews. (www.sherbournemews.com)

To reduce risk of injury or death while working underneath the vehicle, chock the wheels and disable the transmission by removing the gear shift cut out from the fuse board.

This procedure should be performed on a temperate day (60-80˚F). Before you begin make sure that the engine is warm and that the thermostat has opened. You can verify that the thermostat has opened when the

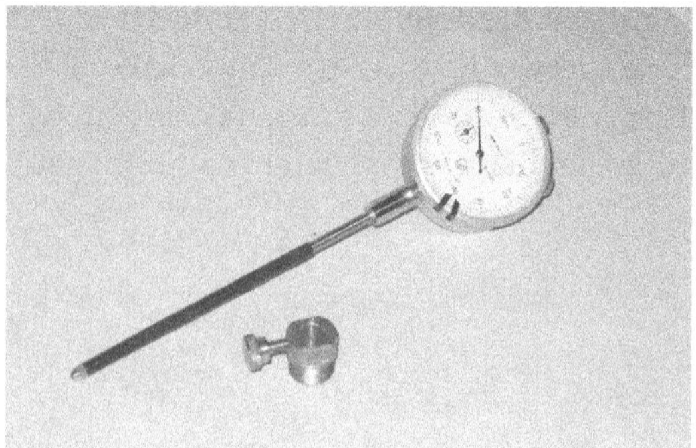

Figure 3-7-1 – Dial indicator, extensions and adapter collar.

Figure 3-7-2 – Volume screw on "A" side carburetor.

Figure 3-7-3 – Dial indicators installed on carburetor suction chambers.

upper radiator hose has become too hot to touch.

Turn each volume screw clockwise all of the way and then turn them counter-clockwise 1-1/2 turns each. Fig. 3-7-2 illustrates the "A" side volume screw. The "B" side is similar.

Unscrew the damper from the top of each carburetor suction chamber. Thread a brass adapter collar into one of the suction chambers and finger tighten. Install one inch and three inch extensions onto each of the dial indicators. Insert each dial indicator into an adapter collar so that the reading is approximately 0.000 +/- 0.020 inches. Unlock and rotate the moveable dial face on each dial indicator so that it reads 0.000 exactly. (See fig. 3-7-3.)

Start the previously warmed engine and compare the readings on each dial indicator. On the carburetor with the lower reading, turn the volume screw counterclockwise with a screwdriver until the reading rises to match the reading on the other carburetor.

Blip the throttle gently and compare the readings again. If the readings are not within a few thousandths of an inch of each other, repeat the adjustment procedure.

Chapter 4
Engine Ancillaries

Section 4-1
Replace Water Pump

The experts recommend a service interval of ten years regardless of mileage because these pumps tend to fail catastrophically without warning. This section is written for a later Series I Shadow with U.S. Emissions, early and European models vary somewhat in their accessory arrangement but the concept is similar.

Removal

A large number of items need to be removed from the front of the engine to access the pump. Proceed patiently and take notes to aid reassembly. Turn off the battery switch or disconnect battery. This is imperative prior to disconnecting the alternator. Adjacent to the lower hose on the radiator is a drainage tap. If the tap drains too slowly or if you plan to replace the lower radiator hose anyway, disconnect the lower hose at the radiator.

Assuming it is still there, remove the rubber boot on the header tank for the coolant sensor connection. Mark and remove the two attaching wires to the coolant level sensor. Try not to lose these 6BA nuts; they are not a hardware store item.

Remove the header tank by loosening the two hose clamps at the rear of the tank and removing the three 7/16-in. A/F bolts. The front overflow hose can remain attached to the rest of the tank and be removed as a unit. Remove the two upper fan shroud pieces by removing the four 2BA nuts and washers.

Remove the four 1/2-in. A/F nuts that retain the fan assembly to the studs on the driving flange. Carefully remove this assembly without damaging the radiator fins. Be careful not to damage the bi-metal strip on the front of the clutch. When you set it aside, place it bi-metal face down. This will help keep the clutch fluid from leaking out.

If a belt tension meter is not available, make a qualitative note of how all of belts "feel" so that you can get them close when everything is put back together.

Loosen the lower alternator bolt and remove the belt. Mark and disconnect the wiring, then remove the alternator. It is possible to change the pump with the alternator installed, but since it is easy to remove, you might as well remove it to gain some working room. In any case, the belt does need to be removed.

Loosen the two air pump mounting bolts and remove the belt. Loosen the Jubilee clip closest to the back of the pump and then remove the air pump. It is possible to do this project with the air pump installed. However, like the alternator, removal simplifies the job and takes only minutes.

Loosen the two 9/16-in. A/F bolts on the power steering pump so the drive belts slacken. Remove the belt pair that runs from the crankshaft to the power steering pump and then to the air conditioning compressor.

Loosen the belt tensioner for the water pump. IMPORTANT: remove the two countersunk screws that retain the pulley to the end of the driving flange. These are not mentioned in the Workshop Manual and no doubt have frustrated many service technicians as they are not visible without a mirror. (See figs. 4-1-1 and 4-1-2.) The water pump pulley should slide off. This helps in removing the water pump belts. Unfortunately, the pulley often sticks to the nose and studs of the driving flange. The belts should be changed when overhauling the pump anyway, so cut the old belts off if necessary and move on.

Do not use a screwdriver or similar implement to pry between the pulley rear and the pump casing. This can result in damage as shown in fig. 4-1-1. A good salvaged pulley can run $275. Spray the nose area of the pulley with PB Blaster or similar penetrating oil. (The "nose" is the shiny face of the pulley in fig. 4-1-1.)

A small gear puller from Sears (#00946905000) will complete this procedure safely. Unfortunately, the drive screw on this model is too long for this application and interferes with the radiator core. Shortening the drive screw by half solves this problem but requires 30 minutes of hack sawing to cut through the case hardened material. (See fig. 4-1-2.)

With the gear puller installed with a small piece of heavy gauge steel sandwiched between the drive screw and the threaded studs, and the fingers grabbing behind the belt slots, apply as much pressure as you can with a box end wrench in one hand while steadying the freewheeling water pump with the other. This is probably no more than 30 ft.-lbs. Do not nick any of the surfaces that come into contact with the belts.

Because the small piece of steel has some spring to it, it will continue to apply pressure to the joint even as the pulley moves.

Figure 4-1-1 – New and old water pump pulleys.

Figure 4-1-2 – Gear puller and elusive countersunk screws.

With the pulley off, remove one 1/2-in. A/F and five 7/16-in. A/F screws. At this point, the only thing holding the water pump to its casing are two dowel pins and residual crud on the mounting face. Because of the geometry of the dowel pins, the pump needs to be removed as close to horizontal as possible. A light rap with a pry bar and brass hammer at several places along the seam should set things free. Be careful not to mar the mating faces with the pry bar.

Rebuild

Take your water pump along with the Crewe rebuild kit (RH10166) to a shop with the right tools and experience.

The water pump shown in this section was not previously rebuilt to factory spec. For starters, it had a non-standard locating screw. It was an Allen type which limited the amount of torque that could be applied for removal. Because high strength Loctite

Figure 4-1-3 – Torch loosens non-standard set screw.

Fig. 4-1-5 – New sealing gland assembly (l) vs. old (r).

was used during assembly, an acetylene torch was used to improve chances of loosening the screw. (See fig. 4-1-3.) Clearance of the driving flange to the bearing housing is carefully noted. (See fig. 4-1-4.)

The old and new sealing gland assemblies shown in fig. 4-1-5 are very different. When the water pump pieces and the kit are all to specification, there is a press fit of one to two thousandths of an inch on the pump bore and interference fit of 12 to 13 thousandths on the impeller bore. Although there is nothing inferior about the domestic rebuild kit previously used, the diameters of its components are slightly larger than the corresponding Crewe items. Loctite was used as well, so extraordinary force was needed to disassemble the pump. In this case, a 25 ton press was used. (See figs. 4-1-6 and 4-1-7.)

Figure 4-1-4 – Clearance between driving flange and bearing housing noted with gauge block (resting on the rear of the vise).

Figure 4-1-6 – Removal of the impeller.

Figure 4-1-8 – Part cleaning in the media blaster.

Figure 4-1-7 – Removal of the bearing pack and sealing assembly.

Figure 4-1-9 – New sealing gland is pressed in.

A media blaster can be used to remove the Loctite and other residue. (See fig. 4-1-8.) The new sealing gland assembly is coated with Permatex Bearing Mount and pressed into place on the back of the pump bore. (See fig. 4-1-9.)

The new shaft and bearing assembly is marked with a permanent marker and straight edge to simplify placement of the locating screw recess in the bore. (See fig. 4-1-10.) Because the bores had been stretched to a light press fit on the pump bore and the impeller in this example, Loctite 271 High Strength Retaining Compound is used as a safeguard. Driving flange and impeller are pressed into place. Take extraordinary care to ensure the proper clearance from the driving flange to the pump housing. (See figs. 4-1-11 and 4-1-12.)

Reassembly

Reconnect the lower radiator hose. With all of the accessories removed, replacing this hose will never get any easier

Figure 4-1-10 – Mark new shaft and bearing assembly.

Figure 4-1-12 – Use gauge block while pressing on flange.

Figure 4-1-11 – Impeller pressed into place.

2 BA	48 to 60 in.-lb.
7/16 in. A/F	8 to 10 ft.-lb.
1/2 in. A/F	16 to 18 ft.-lb.
9/16 in. A/F	29 to 32 ft.-lb.
11/16 in. A/F	42 to 45 ft.-lb.

A/F = dimension across the flats.

Figure 4-1-13 – Torque Specifications for Standard Parts

than at this moment. Install the radiator side first. Because there is little side to side room, the hose has to be pushed on without twisting. Smearing some water soluble, rubber friendly lubricant such as KY on the inner diameter of the hose simplifies installation.

Remove the old o-ring on the water pump mounting face, clean the entire face and replace with a new o-ring. You will not know if you have a proper seal until you put the car completely together, start the engine and pressurize the system. Make extra sure you take your time on this step or you will be doing this all over again in a few hours. A modicum of grease can be used to hold the new o-ring in the groove.

Locate the pump onto its casing so the dowel pins line up. Use the largest hole that corresponds to a 1/2-in. A/F set screw on the pump as your guide. This hole is located near the top. The dowel pins are designed to locate the pump exactly with very little slop. Some fiddling is required to get it to seat.

Note: do not add washers to the six mounting screws; there is not enough clearance to the pump pulley. Torque the screws to Workshop Manual specification. (See fig. 4-1-13.)

Install new water pump belts on the crank shaft and tensioner first. Then, slide the water pump pulley on to the driving flange, slipping the belts onto the back of the pulley as you go. Seat the pulley and install the two countersunk set screws in the pulley nose.

Using a mirror, locate a pry bar in a location that will pull the tensioner tight but will not cause damage to surrounding areas on the engine. Set the mirror aside and grab a ratchet with extension and 1/2-in. socket to tighten the water pump belts to a level comparable to the way it was prior to disassembly. This can be readjusted later.

Install new belts that connect the power steering and compressor to the crankshaft.

Adjust and tighten the power steering pump to achieve subjectively good tension levels. Torque the fasteners to specification. Install the air pump and alternator (each with new belts). Tension the belts subjectively and torque all fasteners accordingly.

Before reattaching the fan, clean the debris from the back of the radiator to improve the efficiency of the cooling system. A 5-10 degree improvement is possible. This is also a good opportunity to remove the fan from the clutch and repaint the fan with high temperature black paint.

Reinstall the fan followed by the fan shroud and then the header tank. Reconnect the coolant level sensor and tighten all of the hose clamps.

Slowly refill the system with 4-1/4 gallons of fresh coolant. (I recommend a 50:50 mix of Prestone and distilled water.) Measure the amount of old coolant drained away relative to new coolant added to ensure that there is no air trapped inside.

Check to see if the belts are rubbing against anything and that the fan will turn freely. Reconnect the battery and start the engine. Check for leaks continuously until the thermostat opens. (The thermostat is open when the large upper radiator hose suddenly becomes too hot to touch.) Top up the coolant and replace the radiator cap.

Turn the engine off and without burning your hands, check that the belt tension is still acceptable. Monitor your garage floor for coolant leaks over the next few weeks and tighten any leaking joints. Recheck the coolant level when the engine is cool and top up if necessary.

Section 4-2
Replace Power Steering Cooler Hoses

This is an easy fix for a common issue: red automatic transmission fluid dripping from underneath the radiator. You might think it to be the transmission cooler at first, but more often than not, it's the power steering cooler hoses.

The materials needed to fix the leaks are limited to one foot of 3/8-inch I.D. power steering return line and four hose clamps. Name brand hose stock such as Goodyear and stainless hose clamps are available at almost all discount auto centers.

Figure 4-2-1 – Radiator support drowning in fluid.

To remove the grille, remove the two 7/16-in. A/F bolts retaining the grille at the bottom. Open the bonnet to remove the four 7/16-in A/F bolts at the top. The top screws also retain shims of varying thicknesses, so it is important to keep track of the original location of each shim.

As you can see in fig. 4-2-1, the radiator support underneath the power steering cooler can completely fill with fluid. Access to the unit is improved if the deflector panel on top of the radiator is removed. (See fig. 4-2-2.) The remainder of the procedure is simple. Remove the bolts securing the cooler, loosen the hose clamps and pull off the cooler and hose pieces being careful not to spill ATF all over the place.

Figure 4-2-2 – Access improves with the deflector panel out.

After cleaning the cooler and surrounding area, cut two new hoses to match the old ones. Reassemble the cooler using the new hoses and hose clamps. Position the hose clamp screw heads facing

Figure 4-2-3 – New hoses and hose clamps in place.

forward to allow future tightening without the need for grille removal (See fig. 4-2-3.)

Take your Shadow for a brief test drive to ensure that all is leak free before reinstalling the grille. You may also want to take this opportunity to clean the radiator matrix and air conditioning condenser before reattaching the grille.

Section 4-3
Replace the Starter Motor

The Shadow starter motor is a very robust item and typically needs little attention. There are a couple of issues that crop up from time that are not cause for replacement.

One is noisy starting. This almost always involves a worn ring gear on the flex plate since it is much softer than the starter pinion. Replacing the starter will not help and sometimes makes it worse. (Trust me.) Unfortunately, replacing the ring gear

Figure 4-3-1 – This ring gear sounded very bad indeed.

involves removing the transmission, something best left to the experts. (See fig. 4-3-1.)

The second issue is an intermittent no start or slow start condition. Quite often this is a result of motor oil leaking from the valve cover gasket down onto the starter connections. The oil attracts dirt and grime and eventually the connection becomes resistive enough to cause an issue. (See fig. 4-3-2.) To replace the valve cover gaskets, see Section 3-1.

Figure 4-3-2 – Dirty connections can lead to starting issues.

Usually, your starter can be rebuilt. If for some reason your starter does need to be replaced, note that Crewe no longer supplies the original style Lucas motors. Instead, the replacement is a Denso gear reduction starter which was phased in during the late Eighties and will work on any Shadow engine. It has a more modern sound, is much more efficient, offers higher cranking speeds and weighs *significantly* less.

Figure 4-3-3 –British Starters unit (top), Lucas unit and shim.

A company called British Starters can also supply a Denso gear reduction starter with a nose piece of their design. These are available for about $300, costing about half as much as the Crewe part. (See fig. 4-3-3.)

To remove the old starter, first turn off the battery switch in the boot or disconnect the battery. This is extremely important because the starter circuit is designed to carry life-threatening amounts of current to the starter.

Park the car on a four post lift and raise it to a nice working height. It is possible to perform this procedure without a lift, but it is not pleasant.

Remove the heavy and light gauge wire connections with a 5/16 BSF* socket. Remove the wire connected to the spade terminal. Remove the ground connection with a 7/16-in. socket.

Using a 9/16-in. deep well socket to clear the stud, remove the lower mounting screw from the front.

Use a 9/16-in. standard socket to remove the screw at three o'clock from the front. This screw also captures a small piggy backed bracket. Slide the bracket free.

From the back, use a 5/8-in. swivel socket with a super long extension to loosen the screw at the twelve o'clock position. This is a blind operation, but since this is the last fastener, you can leave the screw in the housing while a helper steadies and then removes the motor.

You may find that one or two shims are installed between the starter motor flange and crankcase end face. Since these pieces are to account for variation in manufacture

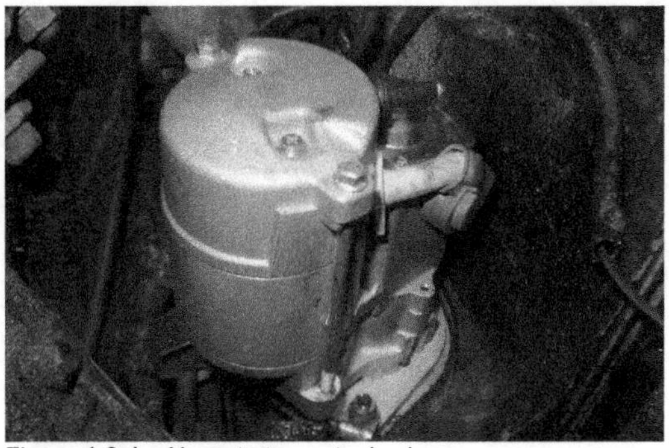

Figure 4-3-4 – New starter motor in place.

of the engine rather than the starter, be sure to reuse these pieces when you install the new starter motor.

If you are using a British Starters unit, the nose piece can be rotated so that you can locate the terminals in the proper place as well as make the three and six o'clock mounting holes accessible.

If the new starter motor has a male spade connection, remove the spade adapter so that the vehicle harness is female.

If installing the smaller type starter, it is likely that you will need to remove the first three brackets supporting the heavy 12V cable so that it can be positioned to hook to the new starter. You will also need to reorient the bend in the cable end in order to align it with the post on the new starter.

Ask your helper to hold the new starter in place along with any shims (also called "packing pieces") while you hold the twelve o'clock screw in place with the long socket extension and swivel socket. Once that screw is in, thread in the remaining fasteners—don't forget the little bracket that attaches on the three o'clock screw.

Attach the ground strap along with the heavy and light gauge connections. Ensure that all of the positive connections are well away from any possible ground paths. Turn on the battery switch or reconnect the battery.

---

* This is one exception to the overall fastener scheme for the Silver Shadow. British Standard Fine or BSF, is a fine threaded fastener from the Whitworth family used extensively on pre-WWII Rolls-Royce and Bentleys. If you do not have a 5/16 BSF (1/4 BSW) socket, try a 13 or 14mm six point metric socket.

Section 4-4
Find Cruise Control Vacuum Leaks

Silver Shadow cruise control issues are almost always the result of vacuum leaks external to the mechanism. This is fortunate, because servicing the cruise control mechanism is beyond the scope of this book.

Please allow me to remind you that this unit interacts with the throttle mechanism of your vehicle. Great care should be exercised to avoid the creation of a hazardous situation. If you do not feel comfortable working on the cruise control system, please leave it to the professionals. Remember too, that a properly maintained braking system on almost any post-WWII vehicle of any type is more than capable of stopping itself (at a reduced rate) even with the throttle fully open.

The system consists of a small electrical panel in the passenger cabin, a control unit located above the "A" side exhaust manifold (See fig. 4-4-1) and a bellows unit attached to the throttle linkage by a lamp chain (See fig. 4-4-2).

Figure 4-4-1 – Control unit over "A" side exhaust manifold.

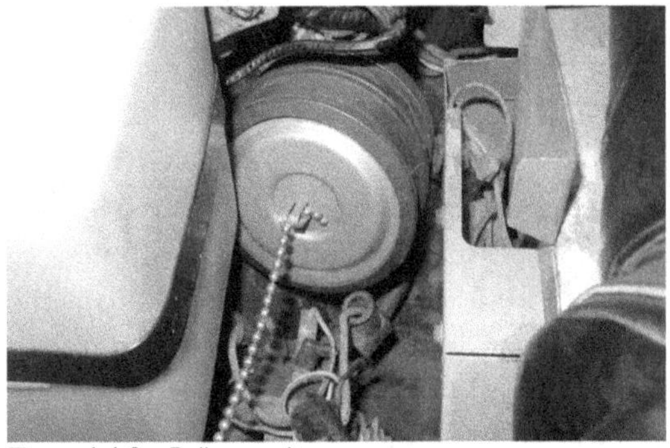

Figure 4-4-2 – Bellows unit.

The control unit has two rotating speedometer cables attached to it: one from the transmission and one to the speedometer in the passenger cabin. There are also two large vacuum lines attached to the control unit: one is attached to the vacuum source (the intake manifold) and the other to the bellows unit.

When the control unit is switched on and set, the unit compares the vehicle speed (from the speedometer cable) to the set speed. If the vehicle speed is less than the set speed, the unit applies engine vacuum to the bellows unit, causing it to contract. This in turn pulls on the lamp chain which opens the carburetor throttles and increases the vehicle speed. Conversely, if the vehicle speed is greater than the set speed, the control unit relieves the unit to atmosphere, which causes the bellows to expand.

If the cruise control fails to operate, fails to maintain a set speed or hunts noticeably around the set speed, investigate the vacuum side first. There are four basic items to check: hose connections, hose integrity, bellows connections and bellows integrity.

The hoses become brittle and tend to take a set as they age, leaking where they attach to the metal tubes on the bellows unit, control unit and intake manifold. If the hoses have been disturbed, old ones can crack quite easily. For vacuum hose replacement recommendations, see Section 3-4.

Similarly, the bellows can become inflexible with age and crack. They also loose their grip on the metal end caps. This can sometimes be rectified with a large hose clamp if the rest of the bellows appears to be in good shape. Finally, the end caps themselves can corrode, creating a rough sealing surface. This can be rectified by sanding or media blasting the caps, then replating them. The end caps appear to have been originally plated with zinc dichromate if you care to restore them to their correct specification. The bellows unit shown in fig. 4-4-2 is painted.

Section 4-5
Replace Air Pump Check Valves and Hoses

The air pump is designed to inject air into the exhaust stream at idle speeds to burn unburned hydrocarbons normally associated with a cold start. In catalyst equipped cars, it also supplies the oxygen necessary for the catalysts to regenerate. The pump represents very little parasitic loss, so why not keep the system running properly and give Mother Earth a little bit of a break?

If your Shadow is equipped with an air pump, as most with chassis numbers higher than 6000 are, the air pump components are normally removed to replace the thermostat. Consider killing two birds with one stone: see Section 2-2.

First, some nomenclature: the *air pump* is driven by a single belt and sits adjacent to the air conditioning compressor. It has a hose exiting from the side or rear connected to a cast *four way connection.* Two of the hoses exit left and right and are connected to

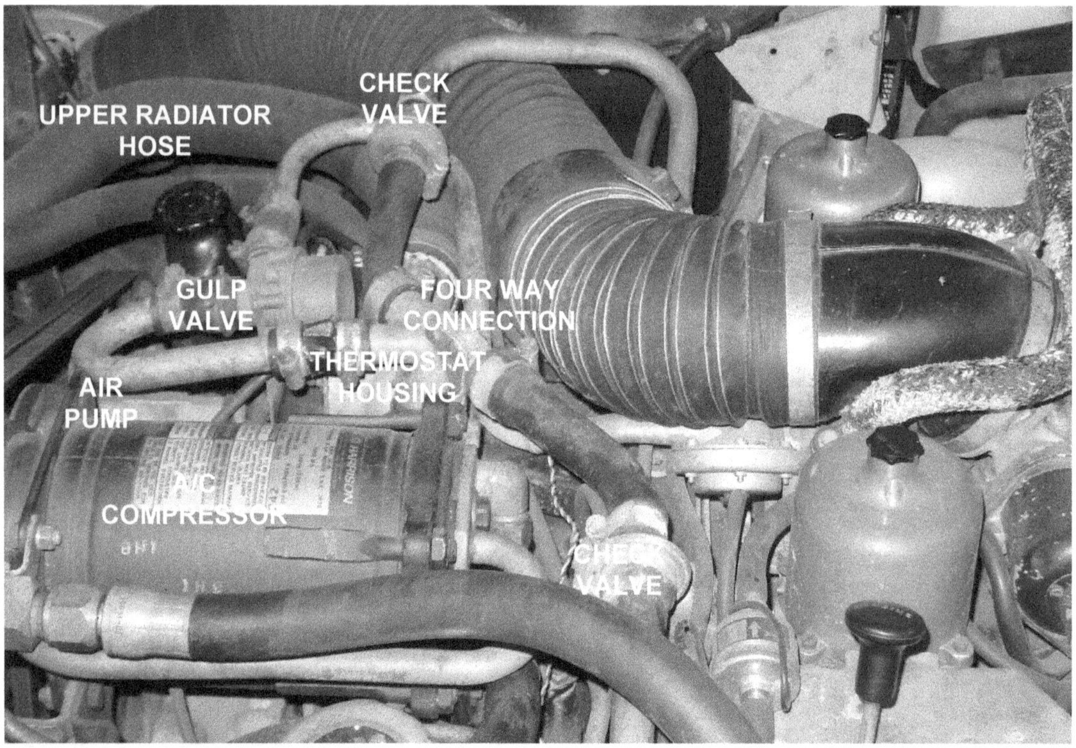

Figure 4-5-1 – Component identification.

check valves (UE74483) which lead to the exhaust manifolds via the *air manifold assembly* (a.k.a. "antlers"). The remaining hose from the four way connection is connected, via a trombone slide shaped pipe and a sometimes a pressure regulator, to the *gulp valve.* (See fig. 4-5-1.)

The elements of the air pump system vary over the model run. The details in this section are specific to chassis numbers from approximately 14300 to 20700, but the general idea is the same.

To begin, remove the trombone shaped pipe and hose that connects to the four way connection together after loosening the hose clamps at each end. Release the four way connection from its bracket by removing the two 2BA screws.

Release the hose clamps at each check valve. Release the hose clamp at the air pump itself. Lift out the four way connection together with the hoses leading to the check valves and air pump.

Remove the two slotted screws that hold the gulp valve to its bracket. (See fig. 4-5-2.)

To remove the check valves, apply a penetrating oil such as PB Blaster to the threaded joint and if need be, apply heat with a propane torch taking care to shield any nearby wiring with aluminum foil.

Removing stubborn check valves calls for the right tools, not channel locks, vise grips or adjustable wrenches. To do it right, use two 1-in. open end wrenches. This will allow you to apply quite a bit of torque without stressing the delicate air manifold assemblies. (See fig. 4-5-3.)

The Crewe supplied check valves (UE74483) are selling for about $60 each at the time of this writing. The ACDelco 214-403 is virtually identical and can be purchased for $14 each or less if you get lucky and find them on eBay. (See fig. 4-5-4.)

After cleaning the threads on the check valve ends of the air manifold assemblies

Figure 4-5-2 – Remove gulp valve from bracket.

Figure 4-5-3 – Remove check valves with the right tools

Figure 4-5-4 – New ACDelco check valve vs. original.

with a wire brush install the new check valves, apply anti-seize compound and tighten to avoid air leaks.

If you intend to replace the air hoses you can follow one of two routes. The proper route: most of the hoses are formed parts and all are available from your Crewe parts provider. The penurious route: purchase 5/8-in. and 3/4-in. high quality heater hose from your local auto parts retailer (such as Goodyear) to duplicate all of the original hose connections. The majority of the connections will use the 3/4-in. size except for the gulp valve connection to the manifold which uses 5/8-in. The latter approach will work just fine but may not pass muster on the judging field.

If you need to replace any hose clamps, remember that all hose clamps are not created equally. As mentioned in the introduction, Jubilee clips are English for hose clamps, except that the band is non-perforated. Figure 4-5-5 illustrates this difference, which may be of concern if you choose to have your car judged.

Build up the entire assembly of hoses, four way connection, gulp valve and associated pipes on your work bench. Install the assembly on the engine, starting with the air pump connection followed by the one way valves and gulp valve. Next attach the four way connection and gulp valve to their respective brackets. You may need to loosen and retighten certain hose clamps to achieve this.

Once you have everything in place and the engine is ready to be started, i.e., you have replaced the coolant if you chose to change the thermostat as well (Section 2-2), start the engine and check for air leaks by listening and also running your hand over all of the connections. Take care to keep your hands clear of any the moving parts.

Figure 4-5-5 – Domestic hose clamp (l).vs. Jubilee clip (r).

Section 4-6
Replace Pressure Relief Valve

The pressure relief valve for the cooling system is located on the right hand side of the header tank under a brass plate.

In the Silver Shadow series, these valves tend to fail in the closed or "stuck shut" condition. If the vehicle overheats, the pressure will continue to build until a radiator hose detaches or splits, leading to a sudden loss of coolant. Considering the importance of the pressure relief valve, you should peridodically verify its operation with a pressure tester. Long time RROC member Dan Docherty made up a pressure tester adapter out of a spare filler cap. I use his. I am certain there are other creative test solutions out there.

At the time of this writing new pressure relief valves are not available from Crewe, but you can buy a second hand one from IntroCar Ltd. along with a new rubber seal (UR15870) and a new gasket (RH12745). There are three different valves to choose from. Most Shadows take the 15 lb. one (UR16730). Your parts dealer will know. You can also compare the number embossed on the old and new parts to make sure you have the right one.

Before you start, ensure that your engine is stone cold or close to it. *If it is not, disassembly of the header tank could result in a sudden boiling of the coolant inside leading to possible injury.*

Remove the five 2BA (11/32-in. A/F) set screws, washers and placard. (See fig. 4-6-

Figure 4-6-2 – Removal of old gasket.

Figure 4-6-1 – Relief valve cover in place.

Figure 4-6-3 – New valve and gasket in place.

1.) Pry the painted brass cover loose from the gasket. Scrape the old gasket off the mating surfaces. (See fig. 4-6-2.) Try to keep the material from falling down into the crevices. If it does, use a Shop-Vac® to pick it up prior to removing the relief valve. If you fail to do this, any remaining material will drop into the coolant and plug small orifices.

The pressure relief valve is held in place by the cover and nothing else. Dislodge the valve and remove the rubber seal nestled underneath. Replace the rubber seal, place the new pressure relief valve with the outlet oriented toward the overflow tube, and install the new gasket. Torque lightly (48 to 60 in.-lb.) in a crosswise fashion as you would a road wheel.

Section 4-7
Replace Viscous Fan Clutch

If the low speed performance of your cooling system has degraded over time, it may be time for a new viscous fan clutch. The viscous clutch is designed to reduce noise and improve fuel economy by decoupling the fan from the water pump pulley if cooling assistance is not needed. Because I have an aircraft quality coolant temperature gauge installed by the previous owner, the dwindling efficiency of my cooling system was easy to spot.

Another way to gauge this is to measure the temperature of the radiator top (not the header tank) with an infrared thermometer after a 20 minute low speed run in summer traffic. The temperature should remain in the 190-215°F range. In addition, note how fast the fan comes to a stop when the engine is switched off. If it turns more than a revolution or so after the engine comes to a stop, you more than likely need a new clutch.

Although this advice applies to all modern cars equipped with viscous fan clutches, the specifics apply to later Shadows with chassis numbers 11111-26700. For this series of motorcars, the Crewe part is UE37061. (See fig. 4-7-1 for a complete listing.)

Ordinarily, I am happy to pay a little bit more for the Crewe part. No need to worry about authenticity or the unintended consequences of fitting parts that are "close enough." In this case, however, the street price for this fan clutch is $360. Obviously, this is not the type of part Crewe would make in house. It is also more than likely that they used their supplier's preexisting design to save on tooling costs.

As it turns out, this part is nearly identical to a $120 Jaguar V-12 part which cross references to a $60 ACDelco unit. ACDelco is not known to cut corners on their specifications in order to sell product in the aftermarket. There are even less expensive no name alternatives, but the best

Fan Clutch Part Numbers

Chassis	Crewe Parts	AC Delco
6300-11110	UE32842	
11111-26700	UE37061	15-4208
30000-42000	UE40167	

Figure 4-7-1 – Fan clutch usage.

Figure 4-7-2 – New unit on left.

bet is the Crewe part or the ACDelco 15-4208 which, incidentally, was used on virtually all V-8 powered GM vehicles throughout the Seventies. (See fig. 4-7-2.)

This entire procedure can be carried out in less than one hour. First, remove the three sets of 7/16-in. A/F bolts, washers and nuts that attach the header tank to the radiator. The header tank does not need to be disconnected from the tubing, just moved out of the way. Remove the two 2BA nuts and washers that retain the upper left hand radiator shroud and remove same. (See fig. 4-7-3.) Remove the four 1/2-in. A/F nuts that attach the fan and clutch assembly to the water pump pulley. The rear face of the clutch tends to rust itself to the front face of the water pump pulley; use PB Blaster or similar penetrating oil to loosen it.

The cooling fan has blades that are spaced asymmetrically for noise reasons. (See fig. 4-7-4.) Rotate the fan so one of the large gaps is facing the left hand bonnet support. While pushing the header tank out of the way somewhat, extract the fan and clutch assembly through the gap created by the missing fan shroud and relocated header tank. The mounting flange of the fan clutch will pass directly in front of the A/C compressor on its way out. (See fig. 3-6-5.)

Remove the four 1/2-in. A/F set screws attaching the fan to the clutch and remove the fan from the old fan clutch. When installing the new clutch to the fan, apply a

Figure 4-7-4 – Asymmetric fan blade spacing.

Figure 4-7-3 – Radiator with upper left fan cowl removed.

Figure 4-7-5 – Fan removal

thin coat of antiseize paste to the face of the clutch and torque the screws to 20 ft.-lb. rather than the higher value found in the Workshop Manual. The combination of short thread engagement and material choice led me to partially strip the threads on a previous installation.

Reassembly is straightforward. Apply a thin coating of antiseize paste to the face of the fan clutch to make disassembly easier next time. Reinstall the fan and fan clutch as a unit. As a final check, be sure to check that the fan will indeed rotate prior to starting the car.

Section 4-8
Replace Radiator

At some point during your custodianship, you may have cause to remove the radiator. Perhaps it is leaking coolant from the braze joints or the efficiency has degraded to an unacceptable point. The Rolls-Royce V8 does not take kindly to overheating; an overheated block tends to lose its main bearing tolerance on the crankshaft, resulting in unacceptably low oil pressure. Setting this straight is no walk in the park. So, an ounce of prevention here is priceless. There is nothing wrong with the procedure outlined in the Workshop Manual, except that it is somewhat more complicated than it needs to be.

Start with a cold engine and slowly open the filler cap to relieve any pressure. Please note that opening the filler cap on a hot engine has the potential to blow scalding coolant all over you, so be careful.

Place a large container underneath the lower radiator hose. Loosen the hose clamp on the radiator side and remove the hose. Coolant will exit at a fantastic rate.

If your car has an integrated transmission cooler (around chassis numbers 13500 and higher), place another pan underneath the two large Bundy tube connections at the bottom of the radiator. Loosen and disconnect with a 3/4-in. crows foot wrench. Transmission fluid will drain from these connections. Blank off the left side tube with a 5/8-in. UNF plug. If you do not blank off this tube, it will eventually drain the transmission entirely.

Loosen all of the remaining hose clamp connections at the top of the radiator: upper, bypass, overflow and header to radiator. Remove the hoses from each connection.

Disconnect the coolant probe boot on the header tank. Quite often these are missing entirely and a pet peeve for many RROC judges. Remove the two 6BA nuts holding the coolant probe wires in place. Bend the tabs that route the wiring around the header tank so that the wires can be moved away from the tank.

Remove the 3 sets of 7/16-in. A/F nuts, bolts and washers that secure the header tank to the radiator. Remove the header tank. Remove the two sets of 2BA nuts and washers securing the upper left cowl to the radiator if so equipped. Using a box end wrench, remove the four 1/2-in. A/F nuts that secure the clutch and fan assembly to the water pump pulley. You should be able to reach all four without need to rotate the

Figure 4-8-1 – Preparing to remove header tank.

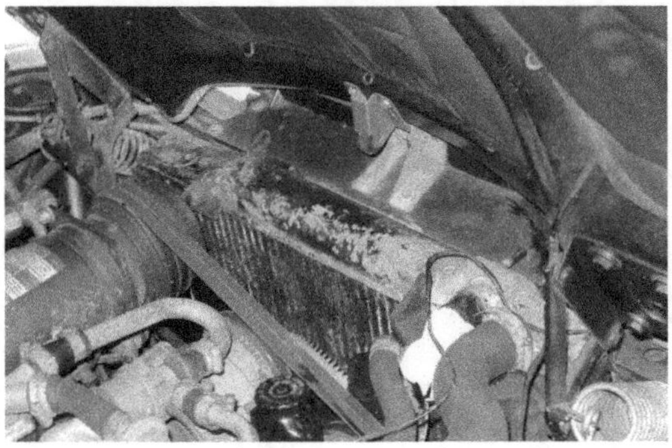

Figure 4-8-2 – Header tank removed.

Figure 4-8-3 – Tired old radiator out on the garage floor.

engine. The rear face of the clutch tends to rust itself to the front face of the water pump pulley; use PB Blaster or similar penetrating oil to loosen it. Remove the clutch and fan assembly taking care to set it on a clean, flat surface with the front face down.

Position a bottle jack to support the bottom center of the radiator and spread some old towels or blankets on top of the entire engine. Place fender protectors on both fenders. The radiator is held in place with four sets of 1/2-in. screws and washers, two on each side, all accessible from the engine compartment. Remove all but one of the upper screws and then locate an assistant. As you remove the last screw, have your assistant hold the radiator in place so it remains balanced on the bottle jack. Lift the radiator from the engine compartment, angling it back to clear the underside of the bonnet. Once it is free, rest it on the towels. Now one of you can get a better grip on it to remove it altogether without damaging the matrix or the fender paint finish.

Since the fan clutch is out, now is a great time to change it if you are in the mood for a side project. (See Section 4-7.) Similarly, the water pump can be changed without removing the radiator but doing so with the radiator out is much easier. (See Section 4-1.) Also, now is a good time to recondition the header tank. (See Section 4-9.)

Reassembly is simple. If the radiator was repainted, ensure that the four mounting screws still thread in. Getting everything to line up is hard enough without having to fight the hardware.

Install the new radiator with the lower cowl in place. For maximum efficiency, there is supposed to be a seal between each of the cowl pieces and the radiator. The lower seal is hard to come by but adhesive-backed foam rubber weather stripping from the hardware store will do well in a pinch. Use the bottle jack to set the height of the radiator and install the top two set screws and washers first.

Reattach the lower radiator hose. This is not the easiest thing to do. The best method I have come up with is to apply a small amount of water soluble, rubber friendly lubricant such as KY on the I.D. of the radiator side of the hose and then install the engine side first. Then, slip on the radiator side. Don't forget to install the hose clamps on the hose *before* you slip it in place!

While you're down there, reattach the transmission cooler lines. Remember to add ATF to the transmission as you will have likely lost more than a quart.

Apply a thin coating of antiseize paste to the face of the fan clutch to make disassembly easier next time. Reinstall the fan and fan clutch as a unit. Reinstall the header tank and then reattach all of the remaining lines including the upper, bypass, header to radiator and overflow.

Slowly refill the system with 4-1/4 gallons of fresh coolant. (I recommend a 50:50 mix of Prestone and distilled water.)

Measure the amount of old coolant drained away relative to new coolant added to ensure that there is no air trapped inside.

Check to see if the belts are rubbing against anything and that the fan will turn freely. Reconnect the battery and start the engine. Check for leaks continuously until the thermostat opens. (The thermostat is open when the large upper radiator hose suddenly becomes too hot to touch.) Top up the coolant and replace the radiator cap.

Turn the engine off and without burning your hands, check that the belt tension is still acceptable. Monitor your garage floor for coolant leaks over the next few weeks and tighten any leaking joints. Recheck the coolant level when the engine is cool and top up if necessary.

Section 4-9
Refurbish Header Tank

The header tank is generally trouble free. If it offers any trouble it is usually a cracked radiator cap or the cap seal, though occasionally the tank itself could be leaking around the clam shell seam. It can also begin to look a bit run down cosmetically after awhile. Given its prominent position in the engine compartment, refurbishing it goes a long way toward improving under hood appearance.

Start with a cold engine and slowly open the filler cap to relieve any pressure. *Opening the filler cap on a hot engine has the potential to blow scalding coolant all over you, so be careful!* Place a large container underneath the drain tap and drain enough coolant to empty the header tank. About two quarts will do.

Disconnect the coolant probe boot, if present, on the header tank. Remove the two 6BA nuts holding the coolant probe wires in place. Bend the tabs that route the wiring around the header tank so the wires can be moved away from the tank.

Remove the 3 sets of 7/16-in. A/F nuts, bolts and washers that secure the header tank to the radiator. Loosen the hose clamps on the two small and one medium sized hose connections to the tank. Remove the header tank.

Use circlip pliers to remove the circlip from the coolant probe. Once the probe is removed, use a small pick to remove the o-ring from the probe body.

Assess the condition of the insulation on the probe leads. If the insulation has deteriorated, remove it and cut a new piece that will terminate at the same point along the probe lead after allowing for shrinkage. Raychem makes a fuel resistant heat shrink tubing (-67 to 302°F), 1/8-in. expanded, 1/16-in. after shrinking, and is readily available from McMaster-Carr. (See fig. 4-9-1.) Replace the probe o-ring, regardless of condition, with a new one (UA 11325).

Remove the pressure relief cover, valve, and seals as described in Section 4-6. The brass cover can be media blasted and powder coated if desired. Caution: the header tank cannot be powder coated as the powder coating process exceeds the melt point of the solder. You can have it media blasted in preparation for painting, though. If you do, be very thorough in cleaning any blasting media out of the tank.

Once the pieces have been refinished, install a new radiator cap seal (UR 5538). Insert the renewed coolant probe into the probe hole. Using circlip pliers, install a

Figure 4-9-1 – Replace damaged insulation with Raychem shrink tubing.

Figure 4-9-2 – Install refurbished coolant probe in tank.

new circlip (UR 13701) and push down on the assembly until it clicks in place. (See fig. 4-9-2.)

Install the header tank with two new small hoses from the auto parts store and new Jubilee clips. As mentioned in the Introduction, Jubilee clips are different than everyday hose clamps because they have a non-perforated band. (See fig. 4-9-3.)

Reconnect the probe wires using the 6BA nuts and washers. Replace the probe boot with a new one as needed. Quite often these probe boots are missing entirely. This is a pet peeve for many RROC judges.

Refill the header tank and replace the filler cap. If the sealing area of the cap is damaged, replace it with a new one.

Start the engine, bring it to operating temperature and check for leaks.

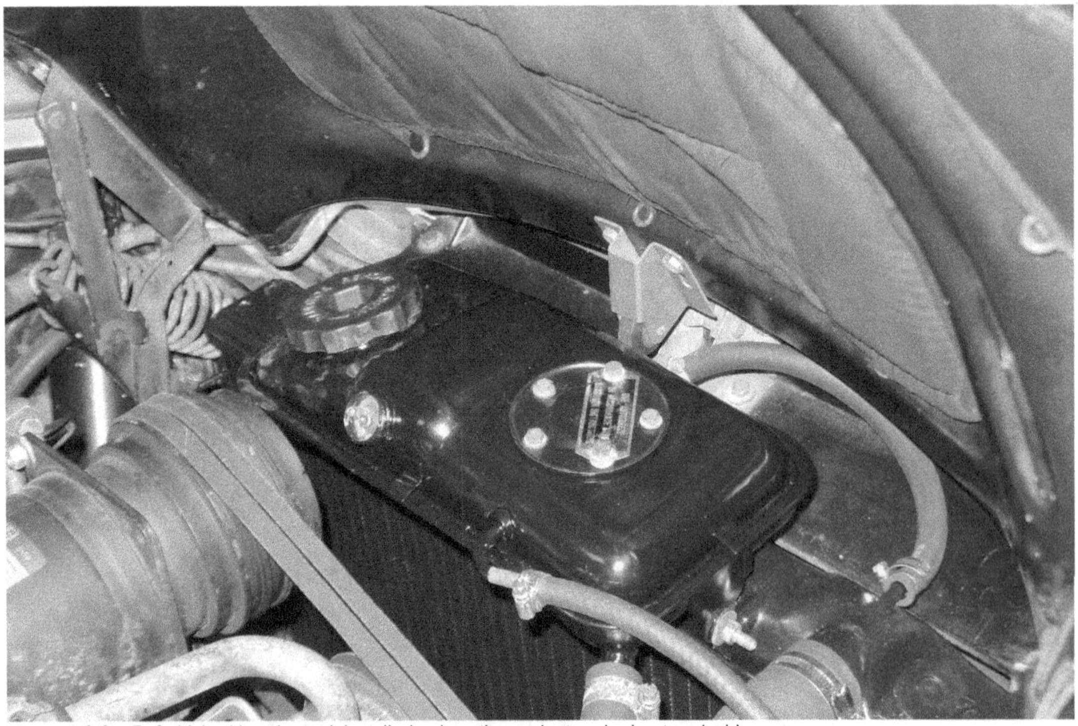

Figure 4-9-3 – Refurbished header tank installed, minus the coolant probe boot and wiring.

Chapter 5
Interior

Section 5-1
Remove Door Panel and Wood Cap

This procedure applies to the four door saloons. The two door coachbuilt variants are similar but not identical.

To remove the door caps, the door panels need to be removed. First, remove the screw covers from the escutcheons on the door panels. Extreme care needs to be exercised because the covers have two cast posts that press into plastic bushings in the escutcheon. If the cover is not removed perpendicular to the door, one or both posts will break. (See Figs. 5-1-1 and 5-1-2.) The Workshop Manual specifies a special tool; you can make one from coat hanger wire. Pound the tip out to a thin blade on an anvil and shape it with a file.

Remove the escutcheons from the door panel. Take care not to lose the small slotted cheese head 3BA fasteners.

On the front doors, slide off the arm rest and unscrew the rail shown in fig. 5-1-3. On rear doors, remove the arm rest by accessing the 2BA set screws from the bottom side of the arm rest shown in fig. 5-1-4.

The door panel is retained by reusable metal clips which are an available Crewe part should any be broken or missing. ($1.69 ea.) Remove the door panel by carefully prying it loose starting at a bottom corner. Carefully mark and disconnect any wiring from the panel.

With the door panels removed, the 2BA set screws retaining the door cap will be exposed. Remove these and disconnect the

Figure 5-1-1 – Escutcheon, cover and cover removal tool.

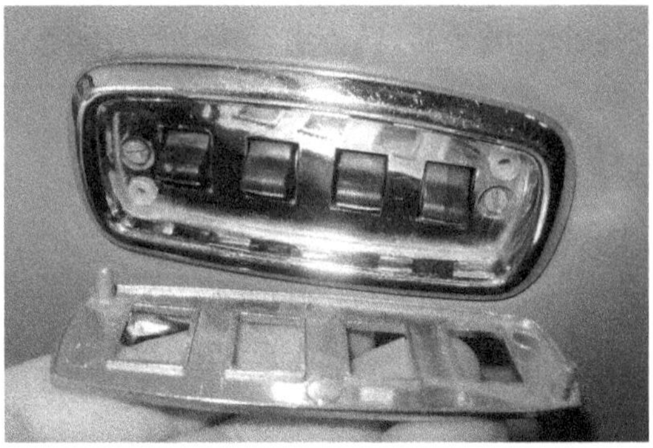

Figures 5-1-2 – Posts are now visible with cover removed.

Figure 5-1-3 – Remove front door armrest to access rail.

lock plunger by loosening the locking nut with a 3BA wrench and then turning the plunger to unthread it.

During reassembly, be careful to set the locked height of the lock plunger so that 1/4-in. remains between the top of the lock rod bushing and the point where the diameter of the lock plunger begins to flare. If set too low, the lock plunger will split the bushing as it moves into the locked position. Fortunately, new bushings and star clips are available.

Figure 5-1-4 – Access rear armrest screws from underside.

Section 5-2
Remove Vanity Mirror

To remove the vanity mirrors, there is one reusable metal clip located above the light switch. This clip is of the same type used on the door panels. There are two tabs at the rear edge of the mirror. With your fingers, carefully pry near the clip location until the front edge swings out. The rear edge can then be disengaged. (See figs. 5-2-1 and 5-2-2.)

Figure 5-2-1 – Vanity mirror clip location.

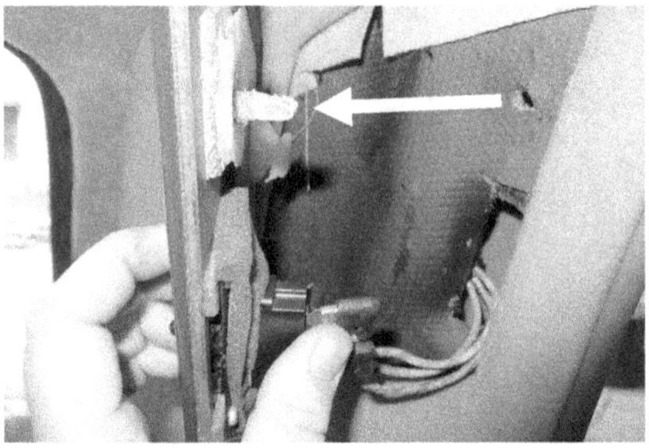

Figure 5-2-2 – Vanity mirror mounting scheme revealed.

Section 5-3
Remove Instrument Panel Wood

To remove the driver's and center section instrument panel wood, unscrew the countersunk screws visible at the top of each panel. Using a 1/16-in. Allen key, remove the ends of the two organ stops. (The "organ stops" are the plungers you move in and out to allow air to come from the "bull's eye" vents in the center of the dash.)

Using your fingers, remove the small pad on top of the steering column and (if equipped) unscrew the cruise control button surrounds.

To remove the glove box wood, open the glove box door and remove the six wood screws from each side. This allows removal of the wood piece but leaves the hinge mechanism in place. Remove either the glove box lamp bulb or fuse #1 to prevent battery discharge.

To remove the wood at the base of the windscreen, the padded cowl or "top roll" needs to be removed first. The cowl is

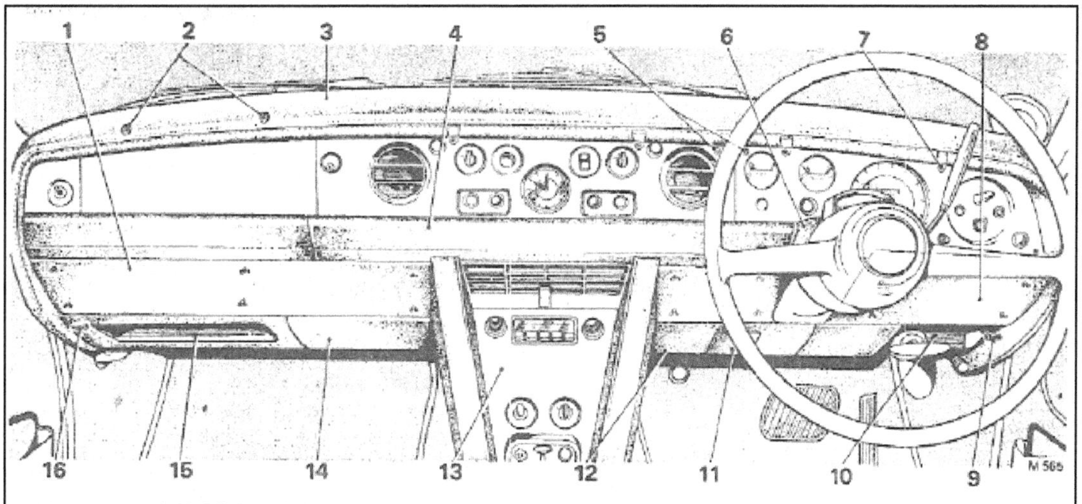

FIG. S108 VIEW OF TYPICAL FACIA TRIM (4-Door Saloon and Long Wheelbase Cars after Car Serial Number 6000)

1 Knee trim pad
2 Nuts securing top roll to instrument panel (2 off)
3 Top roll
4 Lower roll
5 Screws securing polished wood facia
6 Steering column cowl fairing trim
7 Screw—top roll brackets to instrument panel
8 Knee trim pad
9 Screw securing side leg of top roll
10 Handbrake lever trim panel
11 Knee trim pad
12 Trim panel around bonnet release lever
13 Centre console
14 Lower trim panel
15 Parcel shelf
16 Screw securing side leg of top roll

Figure 5-3-1 – Fastening scheme for wood and top roll. Early cars similar.
(Illustration supplied courtesy of Bentley Motors Inc.)

attached to the car body by a 2BA set screw at the bottom of each "ear." It is also attached to the top of the dash by additional 2BA set screws just above the gauges on the instrument panel.

As far as can be determined, the small piece of instrument panel wood surrounding the map light switch is glued in place and is best refinished in place.

The center console piece is retained by the seat switch ears "A", a tab "B" and the radio or 8-track knobs "C". To remove the center console piece, first remove the radio and cigar lighter knobs. Then pull out the floor mounted ash tray and bend the seat switch assembly back just far enough to pull away the wood up and out.

Figure 5-3-2 – Late Shadow I center console wood removal.

Section 5-4
Wood Refinishing

My formula is loosely based on an article that RROC member Mernon Lollich wrote a number of years ago. The stripping part of the process is reasonably gentle and the lacquering part is straightforward, however, if you have a day job and family commitments, this project will take several months to complete in your spare time.

Remove all of the trim pieces from the wood. Chrome strips, rubber boots, mirrors, plastic bushes, and the like are all reusable and are attached by conventional means. The only odd fastening scheme is the illuminated center console seat belt sign, which is held in place from the back with window glazing.

To remove the finish, spread Savogran SuperStrip® liberally with a natural bristle brush. A natural bristle is important because the stripper will consume a synthetic brush. Use rubber gloves and eye protection. As aggressive as this product is,

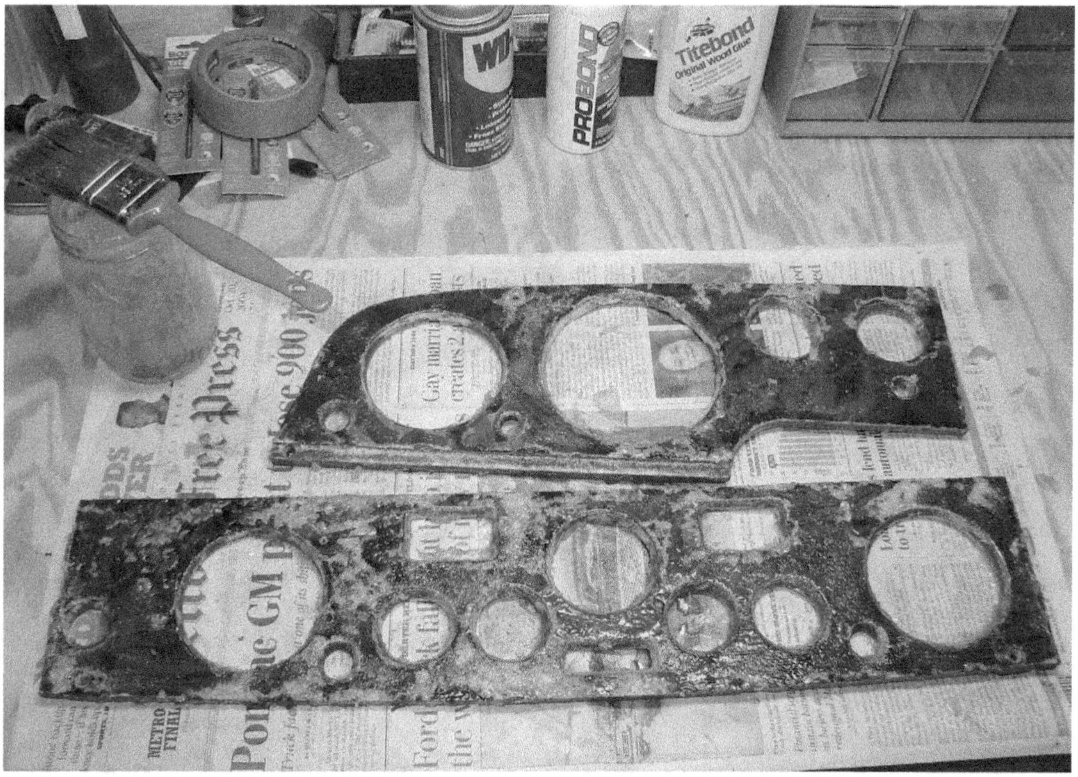

Figure 5-4-1 – Removing finish from wood trim.

it will still take fifteen minutes to begin to see any progress at all on wood with the original finish—a testament to Crewe quality. (See fig. 5-4-1.) Stripping half-baked refinishing jobs takes significantly less time. It is important to stay with the parts to keep applying stripper so that the surface stays wet.

Use Grade 1 steel wool to remove the bubbling finish and then reapply stripper. The most effective batch size is three or four pieces. Once each piece is completely devoid of all finish, wipe it with a lacquer thinner soaked rag. Once each piece is dry, sand it very lightly with 220 grit sandpaper. If the paper becomes clogged, that is an indication that residual finish remains. Repeat as necessary.

Usually you will find the condition of the underlying veneer to be extremely good. It is possible to repair veneer, but this is something best left to the experts.

Sometimes there will be water stains on the work pieces. To address this, apply full strength household bleach to the stain lines using a Q-tip and then wipe the entire piece with a moist rag. Allow each treated piece to dry overnight and sand it lightly with 220 grit sandpaper.

Refinishing the wood without staining tends to look rather blasé. However, it is possible to match the color of the original stain in the majority of Shadows with great results.

Wearing thick rubber gloves, apply Constantines NGR (non grain raising) Light Walnut stain at full strength with a foam pad. Let the wood absorb the stain for 40 seconds and then remove the excess with a clean rag. Then apply Constantines NGR Red Mahogany in the same way, removing any excess after 10 seconds. You can adjust the red to brown ratio by reapplying the appropriate color. If you feel the piece is too dark, Constantines NGR Stain Reducer applied to the part can be used to dilute the

Figure 5-4-2 – Lacquering almost complete.

effect while everything is still wet. It is important to stain all the pieces at the same time to ensure uniformity. The contrast changes considerably once lacquer is applied. Allow the stain to dry overnight. Now begin the tedious process of lacquering in a well ventilated area:

1. Brush on two coats of Deft® Lacquer Sanding Sealer allowing sufficient dry time after each coat. Sand with 400 grit sandpaper.
2. Using Testors™ enamel or similar, paint the instrument apertures brown and the ends of the door caps and glove box door black. After overnight drying, apply a second coat. After drying overnight again, lightly sand the surfaces adjacent to the painted areas to create a perfect transition line.
3. Spray on five thin coats of Deft Clear Wood Finish (Gloss Urethane Lacquer) allowing 30 minutes dry time between coats. Let dry overnight, then wet sand with 600 grit sandpaper. Wipe clean and allow plenty of time for all traces of moisture to evaporate. Any trace of water will cause the lacquered finish to bubble.
4. Spray on two coats of lacquer, allowing 30 minutes dry time between coats. Wet sand with 600 grit sandpaper after the second coat dries overnight. Perform this step five times to add a total of ten coats.
5. Spray on two coats of lacquer, allowing 30 minutes dry time between coats. Wet sand with 1000 grit sandpaper after the second coat dries overnight. Perform this step three times to add a total of six coats. (See fig. 5-4-2.)
6. Spray on two coats of lacquer, allowing 30 minutes dry time between coats. Wet sand with 1500 grit after the second coat dries overnight. Perform this step two times to add a total of four coats.
7. Carefully buff each part using automotive polishing compound. Use Griot's Garage Machine polish (or similar) with an orbital polisher on the pieces you can safely anchor and Griot's hand polish on the fragile pieces such as the vanity mirror frames.

Section 5-5
Fix Slow Window Lifts

Over time, the window lifts seem to move slower and slower. This behavior is not a figment of your imagination. There are two things that contribute to this condition: friction and low voltage.

The models that came after the Silver Shadow, i.e., the Silver Spirit, Silver Spur, Mulsanne, Turbo R, et al., succumb to this issue more readily. This occurs partly because the wire thickness to the window motors was reduced (to save weight or cost) which contributes to a low voltage condition and because the windows themselves are larger and heavier. There is a popular and effective modification to these models that uses the cigar lighter power in conjunction with an added relay in each door. As you will see, this is not necessary on the Silver Shadow series.

To begin, remove the interior door panel from the door with the slow moving window. Refer to Section 5-1 for specific instructions. Be careful here, the trim pieces are easy to damage and fairly expensive to replace.

Friction

The windows that are exercised the least are usually the ones that begin to work slowly or not at all. Regular use of all the windows is usually all that is necessary to keep the lift chains workable and well lubricated.

Figure 5-5-1 – Two areas to access and lube chain.

If the chain is extraordinarily filthy, use brake cleaner to emulsify the dirt. Line the bottom the door panel with shop towels to prevent any brake cleaner from reaching the exterior paint.

Lubricate the lift chain with motorcycle chain lube, such as PJ-1 Black Label®*. Exercise the window lift while applying the chain lube in order to reach as much of the chain as possible. (See fig. 5-5-1.)

Low Voltage

Before proceeding further, a brief explanation of window motor wiring is in order. The following describes the scheme from chassis number 13754 through the end of the Shadow series right through to the

* Motorcycle chain lube is good because it is designed to stick to the chain rather than fly off—the chain stays lubricated longer and the remainder of the door cavity remains clean.

end of the 20,000 series Silver Spirits. Early Shadows are a bit different and also much less likely to exhibit this issue. Referring to the schematic diagram shown in fig. 5-5-2, it becomes apparent that the windows function in a counterintuitive fashion. When no buttons are depressed, both terminals of the motor and window brake are at approximately 12V. Because there is no difference in voltage between the terminals, neither the motor nor the window brake is compelled to move.

Depressing either the master window switch on the driver's door or the individual window switch in the up position grounds one of the terminals on the motor and window brake, causing the window brake to release and the motor to raise the window. Depressing either the master window switch or individual switch in the down position grounds the opposite terminals on the motor and window brake, causing the window brake to release and the motor to raise the window. (The window brake always opens in the presence of voltage regardless of polarity.)

In order for the window to be raised or lowered, a sufficient voltage difference needs to be present across the terminals of the lift motor and window brake. If the voltage difference is less than 10 Volts or so, the motor will have difficulty raising the glass, particularly in colder weather.

Door wiring varies quite a bit between the standard saloons and coachbuilt variants, and from year to year as well. Simply find the two wires from the window brake/motor assembly and measure the voltage at each wire relative to a solid piece of metal on the main part of the car, i.e., chassis ground. (See fig. 5-5-3.)

Without a window switch depressed, this voltage will most likely read very close to battery voltage (approximately 12.8 Volts with the engine off) unless there is blown fuse or open circuit somewhere. While

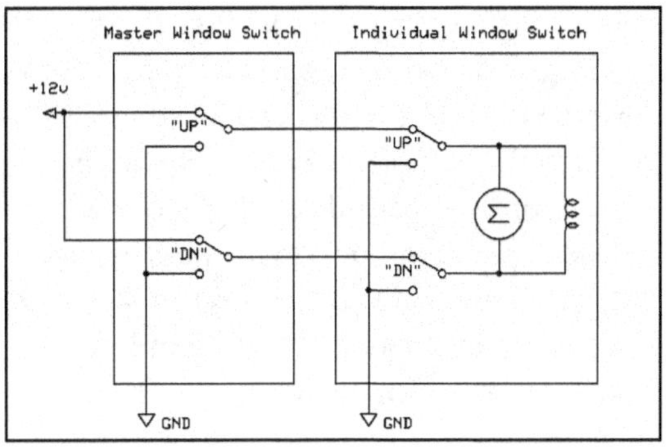

Figure 5-5-2 – Example window schematic.

depressing the window switch, measure the voltage on each side of the window brake/motor assembly, again relative to chassis ground. One reading should be close to the battery voltage and the other should be very close to zero.

If either of these readings deviates from their expected values, that side of the circuit may have a dirty or loose connection somewhere, leading to a voltage drop. Poor connections usually occur in the terminal block of the door itself (found in rear doors only), the kick panels, or the fuse board.

There are two carpeted kick panels. One is located forward of each front door to the outside of the foot well. The carpet is secured by wire brads nailed to a wooden strip located along the leading edge of the door opening.

When tracing faults in the driver's window buttons to windows on the opposite side of the vehicle there are many connections that have the potential to

Figure 5-5-3 – Wires exit the window brake (left) in this example.

contribute to voltage loss in the circuit. Starting with the fuse board, current passes through driver's side kick panel connection through the half of the switch that has *not* been depressed back through the driver's side kick panel through to the passenger's side kick panel and on to the terminal block (found in rear doors only) and window brake/motor assembly on the door being diagnosed.

After the current reaches the motor and brake assembly, the current passes back through the terminal block in the door (if equipped) through both kick panel connections, through the other half of the master window switch in the driver's door that has been depressed, finally reaching chassis ground.

Use the techniques described in Section 9-5 to isolate the problem.

Section 5-6
Refinish Leather

Fellow Rolls-Royce and Bentley Enthusiast Dr. Douglas Dane Wolford has kindly given his permission to reprint his article on leather refinishing originally found in the pages of The Modern Lady. *Although he writes in reference to his well maintained early Silver Spur, the refinishing process for Silver Shadow leather is exactly the same except for a few details identified on the next page.*

Here's a question for you: Have you looked underneath of your front headrests lately? If you do, you might find what I found under mine—beautiful, soft, clean dark leather. This compared to the ugly, somewhat stiff, dirty, and very faded leather surrounding it on the rest of the seat.

My car is a 1983 Silver Spur (NAD-08038) that has been exposed to a great amount of sun in the past two decades. The original leather, which is a very dark red, had faded to almost pink in some places.

After going to the 2002 RROC National Meet and being inspired by the demonstration by the Leatherique representative, I decided to take the plunge and re-dye the leather in my car. The first step was to obtain the products from Leatherique. They are very reasonably priced and the folks on the phone are very pleasant to deal with. Here's my first word of advice: Check the color codes for your car. My code was listed in the Handbook (Owner's Manual).

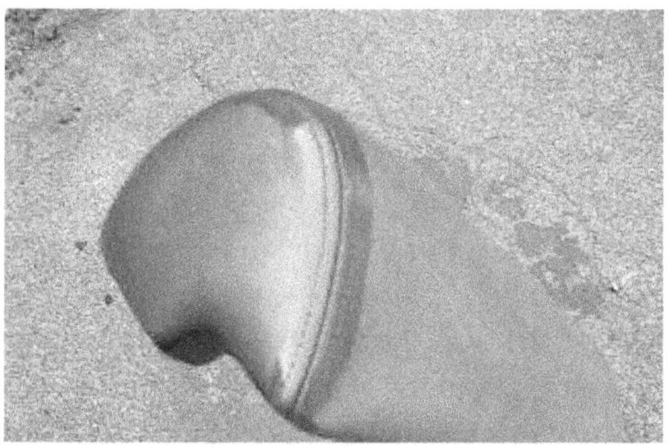

Figure 5-6-1 – Evidence of fading unmistakable.

Once you have your supplies from Leatherique, there are a few other things you'll need to pick up. The dye will need to be mixed slightly before applying, so you'll need a wooden paint stirrer. It is impossible to stir the dye in the bottle it comes in, so I bought a $1.50 plastic pitcher with a lid. I poured all the dye in and stirred it in that. It also gave me a convenient way to pour the dye into a smaller container to paint from, and also resealed so the dye would not dry out. I also picked up a small bowl, some 180 and 300 grit waterproof sandpaper, some acrylic paintbrushes, and a package of cheap terry-cloth towels for spills, cleaning, and clean up. I washed the towels before use, just to make sure there were no contaminants on them from manufacture or shipping.

Once I had all my materials, it was time to start. First step, take all the parts out of the car that you can get out. Both of the front seat bottom cushions and the rear seat cushion can be removed by lifting up at the

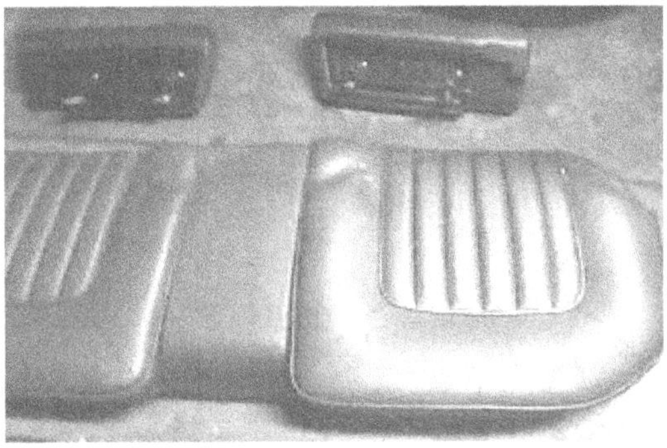

Figure 5-6-2 – Back seat squab removed.

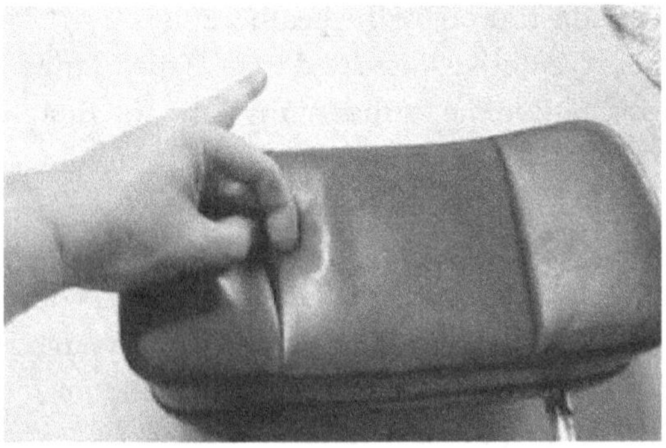

Figure 5-6-3 – Ensure your cleaning is thorough.

front of the seat and pulling the cushion forward and out. The rear seat took some tugging due to its larger size. All four headrests pull straight up and out. The covers over the seatbelt reels on the back parcel shelf can be removed by lifting at the front and pulling up; they are held in by spring clips at the sides and back. The armrests are removed by taking out two Phillips head screws from underneath. This removes the bottom section of the armrest; you then need to remove the bulb from its socket and push the socket up through the piece you removed. I tucked the wires down into the pocket at the bottom of the door. The top half of the armrest is removed with an 8mm socket. I put both the screws and the bolts back into the holes they come out of so I would not lose them, or lose track of where they went.

See Section 5-1 for specifics on how to remove Silver Shadow door hardware, it is slightly different than on the Spirit and Spur.

The rear seatbelts are attached to the body of the car at the bottom by a 17mm bolt, one for each outboard seatbelt. I released both belts and then folded them into the reel area on the rear parcel shelf. The center belt can be folded under the seat. I was unable to remove the rear seatback from the car, so I dyed it in place. Of course all of your sheepskin overlays should be removed. The wool carpet should be removed, especially if the dye and the carpets are of a different color.

The next step is an easy one. Oil your leather. Rejuvenator Oil is easy to apply, and doesn't smell bad. I applied it with a soft sponge to all the leather, both inside the car and the parts I had removed. My leather was 20 years old and I can assume it had not been conditioned in years. It soaked up the Oil as easily as the sponge I was applying it with. Be generous, you will have plenty. I let that set overnight. The next day, I did the same again, more Rejuvenator Oil. It didn't seem to take as much the second day. That set overnight also.

Next step, day three, was to clean all the leather with the Prestine Clean supplied by Leatherique. I cleaned all the leather with Prestine Clean on a terry towel, and then buffed it with a clean towel. Some of the old dye did come off at this point, so be sure and wear good rubber gloves, or your fingers will be the shade of your leather. I left the car undisturbed for 48 hours after cleaning.

Next step, now day five, was to clean the leather. Be sure and wear rubber gloves, as dye WILL come off. Also, be very careful not to let the dye get onto the cloth portions of your interior. I used a mixture of warm water and Prepping Agent. I wiped down a section of leather with this mixture on a towel, and then started gently sanding with the 180 grit sandpaper. I found it was easiest if I kept the sandpaper slightly wet and continued to wipe the leather often.

The old dirt and surface dye will come off as you sand. The appearance is that of the rubber left on the paper after you use an eraser to remove a pencil mark, loose short strands. Be sure and continue to wipe this away. You don't want it on the leather when you start the dye process. After sanding the whole car, I went over everything with Prepping Agent to remove the sanding residue. Then I repeated the whole process using 300 grit sandpaper. Again, I cleaned with Prepping Agent after the second round of sanding. Then I let the car sit undisturbed for 2 days.

I should have been ready to dye at this point, but my back seats still felt sticky to the touch. So I re-sanded the back seats again with 300 grit sandpaper and cleaned again with undiluted Prepping Agent. Again, the car set for 48 hours undisturbed. This was the most difficult part of the whole project; in total I spent approximately eleven hours in sanding and cleaning. Hopefully, your car won't be as bad!

After the car had sat for 48 hours, I then started the leather dying. As stated above, I poured the dye into my pitcher and stirred it with the paint stirrer. Then pour a small amount in a small bowl. The small bowl makes for an easier clean up if you do spill. I used my smallest acrylic brush and started painting on the dye. I did all the seams and crevices of a particular piece first, and then filled in the rest with a larger brush.

The dye will not look like its final color when wet. Mine looked pink when I applied it, but dried to a dark red color.

I started with a couple of small pieces outside of the car to get a feel for how the dye went on. I did the rear seatbelt covers and headrests first. Then I moved into the car and did the rear center console, front console, and one front armrest.

If you are doing two pieces that will touch but need to move, like the front armrests, do one completely and let it dry,

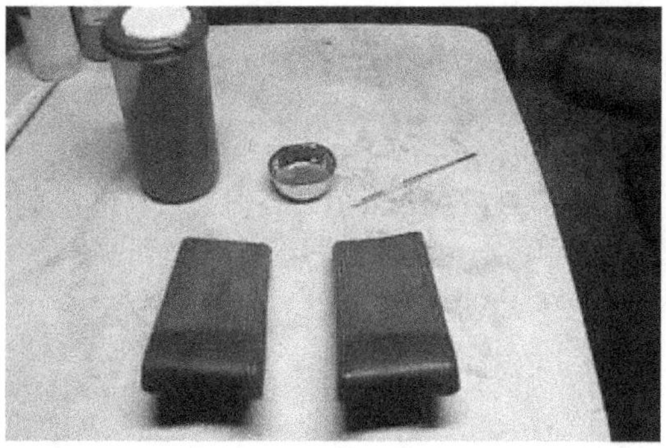

Figure 5-6-4 – Preparing to dye. (No pun intended.)

and then do the other at the end. That way they will not be touching while wet. I did have one armrest touching the center console while it was wet. When it dried and I moved it, it took the dye off of the center console. Those two pieces had to be re-sanded, prepped and dyed after the rest was done.

One trouble spot that can be easily fixed is the leather behind the rear seat, but in front of the parcel shelf. My parcel shelf is cloth covered, and I did not want to get dye on the cloth. I went to a paint store and got an edging tool. It is a foot long metal device with a gentle curve. I placed one edge under the leather, above the cloth, and brushed dye right up against the edging tool. I then slid the tool down a few inches and dyed that section. I did this all the way across. I removed the tool and looked through the back window—no dye on the cloth! I wiped the wet dye off of the tool and used the same process to dye the leather underneath the rear vanity mirrors.

I then went ahead and dyed the rest of the seats in the car, then did the parts outside of the car. The parts outside are much easier as they can be put on a bench or table to work on, and you can move them around. That's why it's important to remove as many parts as you can.

Apply the dye with brush strokes that vary in direction—you want to cover all of the leather and don't want to see brush strokes when it's dry. Also, don't keep your brushes too wet—it makes it easier for brush strokes to show up.

I put two complete coats on my leather. By time you finish with the first, it will be dry enough to do the second. I did both in one marathon session. I spent nine hours applying the dye.

I then let the car sit undisturbed for two days.

After two days, I went back into my garage and was amazed! The leather was a smooth, buff, dark red color. The color was remarkably uniform. This is when I discovered the armrest was touching the center console, so that was redone over the next couple of days.

All that is left to do is to put the car back together. If you put all the screws and bolts into their original holes, then it is an easy matter of placing the armrests back in place and screwing them in. The seatbelts need to be re-bolted to the floor, and the seat cushions and headrests put back into place. It may be difficult to get the seat bottoms back in. They just need some gentle coaxing, and watch your fingers. When they go in, they go all at once (at least mine did). Also, any dye that may have gotten onto the metal rings around the headrest mounting holes, or the metal around the seat and mirror controls can be removed at this point. It comes off easily by gently rubbing with a cloth towel.

All told, I spent about $200 in supplies and chemicals. I put about 30 hours of my actual working time, and a little over a week with the car out of commission. The effort was worth it, without a doubt! My car is never going to look like a concourse car—I drive it too much—but the transformation of the interior has amazed everyone who had seen it previously.

All in all, it is a bit of work, it will take some time, but it is *not* that hard to do. Take your time, be patient! When you're tired, quit for the day. The car will still be there. I think you will be very pleased with the results if you elect to re-dye the leather in your own car.

Section 5-7
Wash Headliner

Long time RROC member and good friend, Bram Over has some quick hints on how to clean a dirty headliner. He wrote this article a few years back when refurbishing the interior of his very early and very handsome Silver Shadow Fixed Head Coupé, which, I might add, once belonged to the Queen's Exchequer.

The internet has all kinds of information: some good, some funny, some useless. These instructions to clean a wool headliner were very useful, I tried it myself and yes, it works. This is how it works:

1. Remove the interior lights and sun visor.
2. Mix a 25% solution of water with carpet cleaner. I used Bissell® Carpet Care.
3. Use a soft 3 x 3 inch brush, and put on rubber gloves. (This solution can irritate the skin.)
4. Have a powerful wet/dry vacuum ready.

Figure 5-7-1 – Wool headliner responds well to cleaning.

Now you are ready to deal with that chocolate brown headliner and 30 years of dirt, smoke, etc. Brush a small area of approximately 12 x 16 inches and immediately thereafter vacuum the moisture up. Repeat the scrubbing with the next small area. You may have to repeat this process a few times. The last time I used long strokes in the same direction to have the "nap" in the same direction. The whole process took me about 8 hours. Some of the water stains did not completely come out, but overall, I was surprised how well it worked and how resilient the wool was.

Chapter 6
Body

Section 6-1
Door Alignment

As the miles go by, the doors tend to walk out at the rear edge so that the door skin is not flush with the rest of the body. To correct this, simply loosen the two large screws on the strike plate and move it inboard.

Dan Docherty showed me the following trick, which will seem obvious after it is described. To loosen and re-tighten the strike plate screws, use a very large flat blade screwdriver that fits the head of the screw. This screwdriver should have a square or hexagonal shaft. Attach a large vise grip pliers to the shaft.

Using a considerable amount of body weight, press the screwdriver squarely into the screw slot and then apply torque using the vise grips. (See fig. 6-1-1.)

Figure 6-1-1 – Tighten strike plate using vise grip pliers.

Section 6-2
Wheel Cover

There is a trick to removing the solid style wheel covers to avoid damage. First, place something on the ground to prevent the cover from being scratched when it comes off. A blanket, towel or your late grandmother's mink coat will all do nicely.

Using the tire valve stem as a reference for 12:00 o'clock, insert your car's tommy bar (tire iron) between the tire and wheel cover at the 1:00 o'clock position. Do not pry the cover; simply pushing it in at an angle that offers the least possibility of bending will ease the locating pin out of the wheel while simultaneously releasing the retaining ring. The cover will then pop off of the wheel.

One reason to remove the wheel cover is to tighten or replace the six clips that secure the trim ring to the wheel cover. Over time, they loosen and cause the ring to rattle against the cover. To tighten the clips, select a suitable socket that is small enough to seat

Figure 6-2-1 – Insert tommy bar at 1:00 o'clock. (See text.)

on the clip but large enough to accommodate the trim ring post. On a padded surface, carefully hammer each clip home.

Eventually, the clips are no longer serviceable. Use a Dremel® tool with a cutting wheel to cut one tooth from the clip and release it from the trim ring. (Please remember to follow the manufacturer's instructions and wear protective eyewear.) New clips are an available Crewe part and are installed using the same socket technique.

Figure 6-2-2 – One of six trim ring retaining clips.

Section 6-3
Polishing Stainless Steel

The first step in polishing stainless steel trim pieces is ensuring that the piece is actually stainless steel. Most grades of stainless are non-magnetic or close to it. A refrigerator magnet can be used to determine whether or not the trim piece is stainless or chrome plated steel. This is important because stainless is very hard and requires rigorous means to bring it to a shine. When applied to chrome, this procedure will ruin the finish.

Once upon a time, this author wrongly assumed that the waist strips were stainless when in fact they were chrome plated. I ruined the finish on three of the eight pieces before I realized this! On a four door Shadow, the following pieces are stainless: grille shell, grille vanes, grille mascot (but not the mascot base), headlight bezels, sill strips, kick strips (inner and outer), rear view mirror surround (but not the base), window frames, wheel covers and B-post covers. However, do not take my word for it. Make sure that the pieces you are about to polish are non-magnetic.

As an aside, polishing chrome plated items is simple: hand rub Simichrome polish sparingly into the part with a cloth until it shines. If this does not work, consider having the part replated. Simichrome® is a German made paste that can be ordered from Eastwood, among others.

A capital equipment purchase is required to polish stainless steel properly. A 1/3-hp Buff Shop Kit (#13310) from Eastwood Co. along with a stand plus a ½-in. yellow treated, ventilated buff might be considered entry level. The kit includes a variety of polishing compounds and buffs for metal polishing plus related safety equipment. (See fig. 6-3-1.)

Please treat the following as a supplement to the instructions that come with your equipment and your own common sense. This is dangerous equipment that turns at high speeds and uses an appreciable amount of electrical power. Be careful! You should wear eye protection, a dust mask and heavy leather or pigskin gloves at a minimum.

Before starting, mount a sisal wheel to one side and a yellow treated, ventilated buff to the other according to the manufacturer's instructions. As you become immersed in your polishing work, take care not to contaminate the polishing wheels with the wrong compound.

Figure 6-3-1 – Basic buffing kit, outdoor use is a plus.

Step 1: Sisal Wheel with Emery Compound

The 1/3-hp Eastwood motor turns at 3600 RPM and comes with six inch diameter polishing wheels. The polishing compounds come in sticks that look like giant crayons. With polishing wheels mounted, touch the emery compound to the sisal wheel for one to two seconds. To start, pick a simple part without complicated features such as the rear view mirror surround.

Applying the part to the wheel below the centerline of the motor shaft, create a uniform haze by moving back and forth in a methodical fashion. Be careful not to present a leading edge of the part to the wheel or it will snatch it from you at approximately 64 mph! Periodically add a bit more emery compound. Too much compound is evident if it begins to build up on the part. Be careful not to overheat the part as it will discolor. With stainless, however, you will notice the heat through your gloves before this becomes critical.

Step 2: Yellow Treated Ventilated Wheel with Stainless Steel Compound

The Buff Shop Kit comes with a spiral sewn wheel which can be used with the stainless compound at this stage. The cutting action of this combination, however is abysmally slow. Instead, use the stainless steel compound with a yellow treated ventilated buff purchased separately.

To eliminate contaminating the wheel with residue from the previous step, clean the part with a rag soaked in lacquer thinner or other similar solvent. In the same manner as Step 1, move the piece back and forth past the wheel. You are removing the peaks in the material created by the previous operation, so best results occur when you alter the buffing path 90 degrees relative to Step 1. Progress with this combination of wheel and compound seems to be very sensitive to the amount of compound present on the wheel: all progress comes to a halt when there is too little and an egregious amount of residue builds up on the piece if there is too much.

In this step, the part transforms from a brushed stainless appearance to mirror like. If it does not look mirror-like, repeat this step with careful attention to the amount of compound used and the angle of the wheel to the part. If deeper scratches are present, step 1 needs to be repeated in selected areas, or the scratches need to be addressed with 320 grit sandpaper.

Step 3: Loose Section Wheel with White Rouge Compound

As before, clean the part to avoid contaminating the next wheel. This wheel deepens the luster and removes any final haze. Two ½-in. wide loose section wheels stacked together seem to work well on this stage. Again, best results are achieved if your buffing path is at a 90 degree angle relative to the previous step.

Buffing Equipment Recommendations

A 1/3-hp motor seems to offer adequate power, but the larger motors can drive 9-in. diameter wheels and have much longer spindles. This allows significantly better flexibility to access all angles of complex parts without striking the motor housing.

The 1/3-hp Eastwood kit includes a complement of 6-in. wheels. In lieu of the kit, the following complement of polishing accessories for a modern Rolls-Royce application (i.e., no brass or aluminum to polish) is recommended:

For 1/3-hp motors, purchase:

- One 8-in. dia., ½-in. sisal wheel
- One 8-in. dia., ½-in. yellow treated ventilated wheel
- Two 8-in. dia., ½-in. thick loose section wheel
- Emery compound
- Stainless steel compound
- White rouge compound

For 1/2 or 3/4-hp motors, purchase:

- One 9-in. dia., ½-in. sisal wheel
- One 9-in. dia., ½-in. yellow treated ventilated wheel
- Two 9-in. dia., ½-in. thick loose section wheels
- Emery compound
- Stainless steel compound
- White rouge compound

Figure 6-3-2 – Die grinder with mushroom shaped buff.

Polishing Complex Shapes

The wheel covers and headlight bezels are troublesome to polish on the buffer stand. One approach is to use a 3-in. mushroom shaped sisal buff for Step 1 and 3-in. mushroom shaped cotton buffs for Steps 2 and 3. Unfortunately, conventional drills spin at 1200 RPM (versus 3600 on the pedestal buffer) and the diameter of the buff is only 3-in. (versus 6-in. or more on the buffer). To get the same polishing speed, the mushroom buff would have to be driven at 7200 RPM or higher.

For those that have access to compressed air, a pneumatic die grinder with a 1/4-in. collet is a rather elegant solution. (See fig. 6-3-2.) The Ingersoll-Rand IR307A shown offers 1/4-hp and draws 3 CFM at 90 psi. And if you have ever seen *American Chopper* on the Discovery Channel, you might convince yourself that you could fabricate a custom motorcycle with your die grinder, some chrome-moly tubing, and a TIG welder!

Section 6-4
Bare Metal Repaint

The following is an account of my personal experience in having my Shadow repainted. A first rate paint job equal to the original is a non-trivial exercise. At 2006 prices and Midwest US labor rates, expect the total cost to exceed $23,000 including parts, labor and plating costs.

Sometime during the spring of 2004, I decided to have my Shadow repainted. Although she was a *Modern Lady* cover car a few years ago, two decades spent in Southern Florida along with a ham-handed Eighties respray made her a "20 footer" by most accounts. Although Mary Richards might have dated the man driving the Sand over Astrakhan Shadow back in 1973, a contemporary eye might describe her livery using scatological terminology.

After canvassing local RROC luminaries, two potential paint shops were identified. Both would be able to start work in the fall of 2004 and estimated four or five months to completion. The first shop had repainted at least four Shadows with which I am familiar. The second shop, 203 Custom Car Works, had never repainted a Shadow, but I was impressed with some of their prewar Rolls-Royce and hot rod fabrication work. As a two man operation, 203 has no everyday collision business to distract them. I found that I had an instant rapport with the owners Paul and Scott, and their shop is located very close to my office. In the end, I took the plunge with 203 despite the fact that Paul and Scott wanted more money than their competition. I do not regret my decision; the rapport and proximity definitely contributed to a very high quality end result.

During the summer, Richard Vaughan and I started brainstorming on a pleasing new color combination that would compliment the medium tan hides with deep fawn carpeting. At a meet at the Gilmore Museum in Kalamazoo, Michigan, I happened upon Peter Heydon's 1989 Corniche in Brewster Green with tan top and hides. (This fabulous automobile now belongs to Jack Pfeifer of Bloomfield Hills, Mich.) Brewster Green was it, but it might appear funereal to paint a Silver Shadow that color in its entirety, so the search for a complementary top color was on. "Perhaps some kind of silver," Richard suggested.

At the Centenary RROC Meet in Monterey, I stumbled on a book of period color chips in the bazaar. Having purchased them, I found the Brewster Green was indeed offered in the chrome bumper Shadow era. More importantly one chip stood out as a fantastic top color: a dark silver chip with the slightest green tint – Pewter.

As summer became fall, I wrote a statement of work for the guys at the paint shop to use as reference. A long job like this might be less stressful for all concerned if I spelled out my requirements prior to any work being done or any money changing hands. (See fig. 6-4-1.)

I also began to amass the myriad replacement seals needed to do a proper job. It does pay to shop around. In my case, I saved about 30% of list price for the Crewe parts that I ordered from IntroCar Ltd. Naturally, I missed a few items and in the end benefiting from the lightning fast service of both Albers and Newport Auto Center. A complete list of parts can be found in fig. 6-4-2.

What follows at the end of this section is a six month photo diary of a very thorough bare metal repaint. Now that I have lived through this experience, there are a number of things learned along the way that might save you time, money or heartache:

- The paint shop quote for disassembly, stripping, prep and paint should exceed 400-450 hours if they are serious about doing a quality job.
- A 2BA tap is very handy for cleaning out paint from the majority of the threaded holes on the body and door panels. British Tools and Fasteners has low cost ones available for less than $10.
- A #6-32 tap is handy for cleaning the small threaded holes in Lucas lamp assemblies. These are available at larger hardware stores.
- The chrome plated brass wiper spindle jam nuts are an odd size (5/8-20) and best treated gently, rather than searching for a replacement. (Fortunately, Acme Car Service had a used set waiting for me.)
- Having the doors completely disassembled is a once in a lifetime opportunity to thoroughly clean and lubricate the window lift chains. (See Section 5-5.)
- Most of the small electrical connectors that join the exterior lamps to the wiring harness are available and reasonably priced Crewe parts. Plan ahead when the trim is removed and order what you need.
- Plan on things going wrong at the end of your restoration, e.g., broken windshield, ill fitting seals, electrical gremlins, etc.
- Do not plan on a major show in same year that your car is supposed to be ready; it stresses everyone out and tempts one to make compromises in the interest of speed.
- Watch the clearance of the front door seals to the B-post sheet metal screws. If the heads protrude too far, they will tear the seal.
- Masking tape, if left long enough, will pull the finish off of Connolly leather.
- Plan to be very involved in the process.

Since I enjoy diving into the technical details and tracking down hard to find parts, restoring the Silver Shadow was definitely more fun than heartache. Elsinore appears as if it were 1973 again. Would I go through this again someday? Absolutely. For the immediate future, though, it will have to be with someone else's money!

Figure 6-4-1 – Example Statement of Work for Paint Shop

Statement of Work
1973 Rolls-Royce Silver Shadow Bare Metal Repaint

Trim and seal removal

- Remove all brightwork including bumpers, front and rear windshield seals, door handles and other exterior ornamentation. (Owner will remove grille prior to arrival at 203 Custom Car Works so that it can be polished.)
- Evaluate condition of bumper pieces and notify owner if rework or replacement is needed.
- Remove non-standard rub strip permanently. These run along the sides of the vehicle below the door handles and, unlike the sill and waistline strips, have rubber inserts.
- Remove door, hood and trunk seals, including seals at base of side window frames.

Bodywork

- Repair corrosion damage to driver's floorpan. Be careful not to disturb linkage geometry in the "rat trap".
- Verify integrity of scuttle drain hose behind trailing edge of left front wheel well.
- Address integrity of front wheel wells.
- Repair buckled right rear wheel arch.
- Address poor panel fit on both rear quarter panels relative to the trunk lid and trailing edge of rear doors.
- Verify that panel gaps on remainder of vehicle are within reason. (Panel gap specs are in Section S of the Workshop Manual.)
- Repair minor dents and dings including dent in top of trunk lid.
- Repair corrosion damage on right side of rear valance.
- Address source of paint blister on left front fender.
- Repair top of door sheet metal so that it properly mates with wood cap.
- Fill holes left by non-standard rub strip.
- The goal is to correct thirty years of second rate repairs, corrosion, and general aging. The expectation is to proceed with this in mind, notifying the owner of additional major items in need of replacement such as sills or quarter panels.

Paint

- Make note of asymmetric distance of pin stripes to waistline strip.
- Note location of two tone body color transition line, including inside door frames.
- Repair leaded joints, prepare and prime body.
- Note that hood, trunk and door skins are aluminum.
- Repaint body including channels in hood and trunk, fuel filler area, door frames, and doors, including side and bottom edges. (Silver Shadows originally assembled BIW and then painted. The goal is to approximate that approach.)
- Repaint wheel trim rings in bottom color. (See note below.)

Figure 6-4-1 (continued)

Top Color: Pewter Metallic
Rolls-Royce No. 9510371
Orig. mfg.: Inmont
Current mfg.: Glasurit

Bottom Color: Brewster Green
Rolls-Royce No. 9510393
Orig. mfg.: ICI
Current mfg.: PPG
PPG Code: 44525

Pin stripe: Ivory

- Owner will gladly approve paint color prior to application if that is desired.
- Sand, clear coat, buff, etc.
- Pin stripe body as it was prior to paint removal.
- Wheel trim rings are pin striped as well, owner to provide photograph and/or measurements.

Reassembly

- Reassemble car with new seals, o-rings and clips, unless otherwise noted.
- Owner will provide polished grille for 203 Custom Car Works to install.

Notes

- If work requires complete removal of rear windshield, please permanently remove aftermarket third brake light.
- Spare tire is located outside underneath trunk. Please remove prior to painting if overspray is a concern.
- The owner will provide a CD-ROM of Workshop Manual and Parts Manual plus printed portion of body section. Among other things, these documents provide invaluable aid in terms of proper seal and trim installation, panel gaps plus blow up diagrams of everything.

Figure 6-4-1 (continued)

- Jacking, lifting and towing: the rear subframe to which the differential is attached looks like a bridge truss, but it is actually fairly fragile. The car currently has a brand new one because of damage caused by previous owners. The replacement part was over $2,000. Please do not apply stress to this part in any way!

- The car will be delivered to 203 Custom Car Works with the grille and inside door trim removed. Wheel trim rings will be removed and stored in the trunk.

- The intention is to provide a brand new seal, o-ring, gasket or clip for every exterior piece that you remove. Please advise the owner if there is a shortage. See below for parts list. Exceptions: the secondary seals that run along the outside of the window frame portion of the door are to be removed but not replaced. The large rubber seals under the stainless kick plates can be reused.

- The rear valance consists of three pieces: the valance itself and brackets that create a subtle tuck in the trailing edge of each quarter panel. Please consult the owner if these brackets are missing.

- Silver Shadows have an engine driven high pressure servo-hydraulic braking and height control system. (It is not a vacuum assisted master cylinder system.) As such, they get really persnickety if they are not started at least every two or three weeks. So please do. (As with most cars of this era, set the high idle by pressing and releasing the accelerator prior to starting. Tapping the accelerator will bump the idle down once it is warm.) When you do start it, ensure that both of the brake pressure warning lights in the center of the dash go out prior moving the gear selector out of park. As mentioned above, the confluence of all of this hydraulic pipe work and brake pedal linkage is immediately behind the driver's floor pan area in the "rat trap". If you need to disturb this area (you should not need to) consult the owner.

- There is a battery cut off switch adjacent to the battery on the left side of the trunk.

- All fasteners less than 3/8-in. in size are British Association type. Both the wrench sizes and thread pitch are entirely different than normal SAE or metric fasteners. The owner has a source for British Association open end wrenches and ¼-in. drive sockets and replacement fasteners if needed. (All fasteners 3/8-in. and greater are typically SAE.)

Figure 6-4-2 – Parts List for 1973 Silver Shadow

Qty.	Part No.	Description
90	UB12419	Female body side clip, for waistline, sill and bonnet strips
60	UB12416	Male clip, for waistline strips
20	UB12418	Male clip, for sill strips
10	SPC2182	Male clip, for bonnet strip
1	UB19116	Glass, remotely operated door mirror
2	UB35762	Seal, grille to body
2	UB12889	Seal, twin headlamp to body
2	CD3116	Seal, front indicator lamp to body
2	UB16425	Hood/bonnet beading
1	UB43421	Seal, main bonnet
1	UB12467	Seal, windscreen
2	UB31534/5	Seal, front door main
1	UB31536	Seal, right rear hand door main (swb)
1	UB31537	Seal, left rear hand door main (swb)
1	UB12377	Seal, left front hand upper door hinge
1	UB12378	Seal, right front hand upper door hinge
2	UB10626	Seal, front lower door hinge
2	UB10625	Seal, rear upper door hinge
2	UB10677	Seal, rear lower door hinge
2	UB13438	Seal, sill jacking point
1	UB12408	Seal, rear windscreen
1	UB15695	Seal, main trunk
1	UB14153	Seal, number plate (USA)
1	UB13854	Seal, boot badge
8	UR8130	Seal, over-rider to bumper
1	UB12119	O-ring seal, trunk handle
2	CD4858	Seal, side flasher lamp center
4	CD5385	Seal, side marker lamp front and rear
2	CD5141	Seal, rear tail lamp to body
1	CD5742	Lens, rear stop lamp – amber (USA)
2	UB12360	Seal, bumper bracket to lower rear panel
2	UR15450	Conical seal, rear quarter bumper to fender wing
2	UB12889	Headlamp door seal
4	UB1561	Air Grille Seal
2	TY341	License Plate Lens
2	CWB301	Connector - Double
2	UB16220	Rear door conduit
4	UB17748	Window frame seal
4	UB12208	Window wiper
2	UD11118	Wiper shaft seal
4	UD11630	Headlamp seal

November 27, 2004

Car is delivered to 203 Custom Car Works in Ypsilanti, Michigan. Door panels, wood, wheel covers, grille, and kick plates have been removed.

December 5, 2004

Some brightwork has been removed...

This right front fender shows a "biopsy" performed to assess paint thickness and underlying condition. Paul reports, "all we found so far is a shit load of paint." Note that the trim was not removed for the respray back in the Eighties.

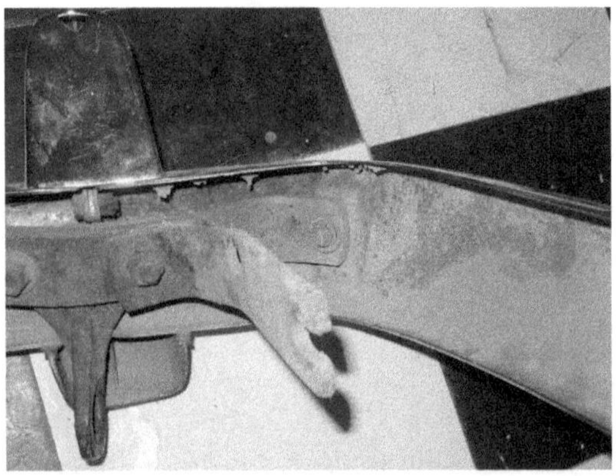

December 8, 2004

Moderately corroded bumper assemblies to be disassembled and sent to D&D Classic Bright Works in Piqua, Ohio for replating. Bumper brackets will be powder coated locally at Dun-Right / Q.C. Coatings in Shelby Township, Michigan.

January 3, 2005

Paint removal has begun in earnest. Bonnet and boot lid have been removed...

Accident repair with large amounts of body filler revealed on left rear quarter panel.

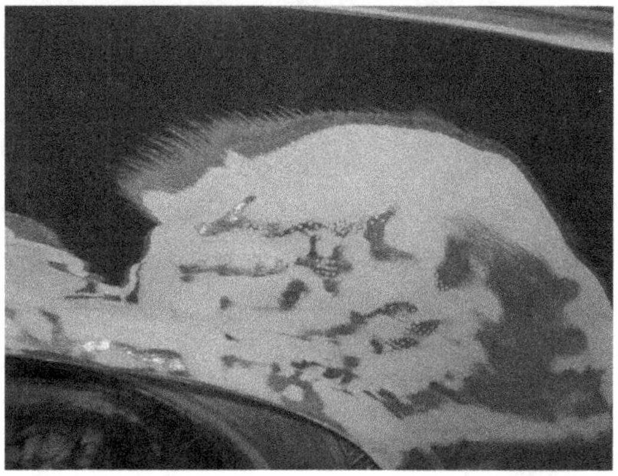

Accident repair with ludicrous amounts of body filler on right rear quarter panel becomes visible. The white stripe that starts at the bottom of the fuel filler door and rises to the right is almost 3/8" deep with filler. Discussion begins on whether we should replace both rear quarter panels...

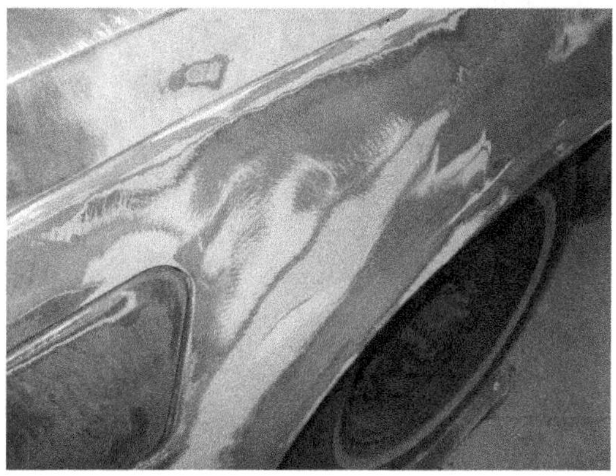

Roof has been chemically stripped but not yet sanded down.

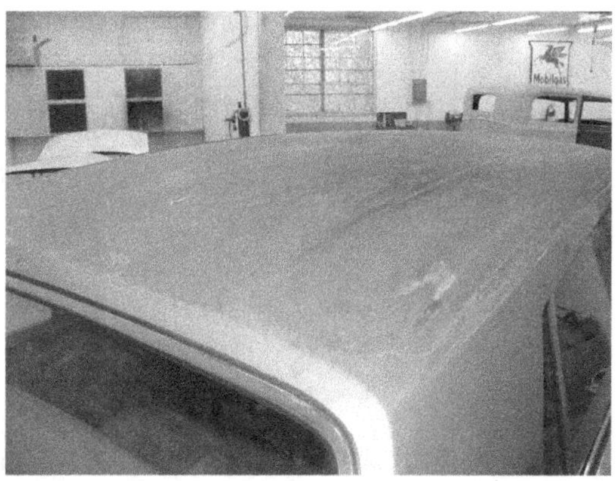

January 3, 2005 continued...

Bonnet and boot have been completely stripped.

January 7, 2005

All doors, windshield, scuttle and miscellaneous trim have been removed...

Close-up on door apertures: I was concerned that there might be rust behind the stainless steel cover on the B-post, but apparently I was imagining things...

Rear doors have been stripped of stainless frames, glass and window lift mechanisms...

Rear glass has been removed. New quarter panels are $1,500-$2,500 each. Further, attaching a new quarter panel with proper alignment at all attachment points will be just as much work as removing all of the body filler and working these back into shape. Decision is made to fix what is there.

February 18, 2005

All four doors primed.

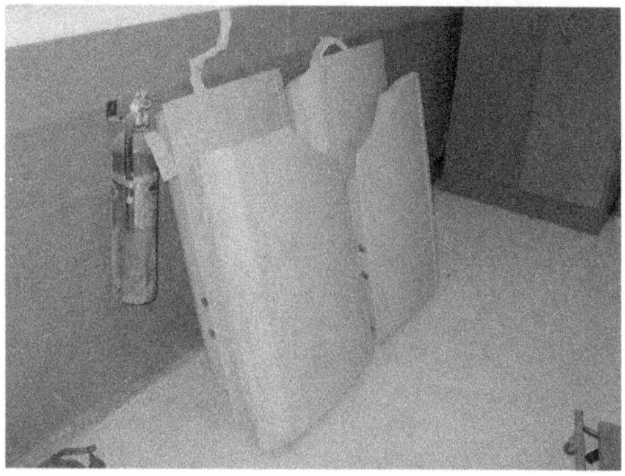

Underside of boot lid. Carpet was destroyed in the process of removal. A replacement will need to be ordered from Hilborn. Progress has slowed on the Shadow in favor of a DeSoto whose brake drum appears at the lower left...

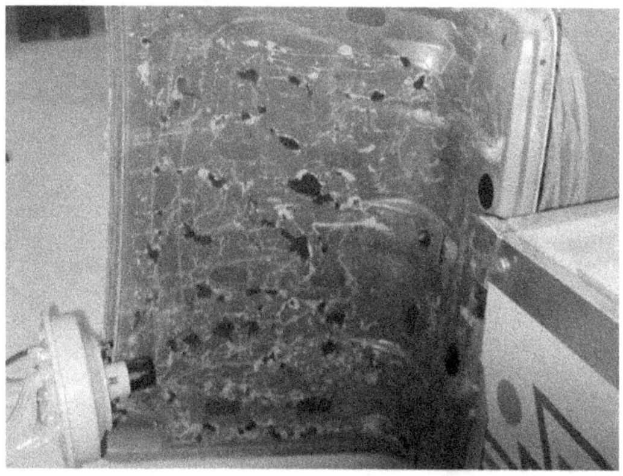

The good news is that the bumpers are back - plated and polished to Hubble perfection...

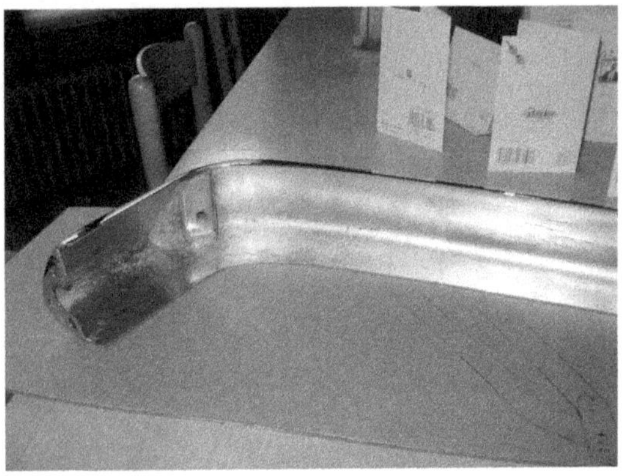

The reverse side looks great, too. All traces of rust are gone...

The powder coated bumper brackets are back also and look equally good.

March 15, 2005

Ninety-nine percent of the filler material has been removed from this quarter panel and it ready to be assaulted with a hammer and dolly...

All filler material is gone from this quarter panel as well and is also ready for a beating. The edge where the lower lip of the boot lid meets the rear valance is pure Bondo and a replacement panel may be in order. We will know soon enough...

This is the right front quarter with the splash shields removed. The left front splash shields are damaged and need to be repaired or replaced...

I am beginning to believe I am getting my money's worth...

Door jambs are stripped...

The part of the rear valance below the right hand tail lamp has some rust through. This is the only rust found on the car other than the driver's floor pan.

March 30, 2005

Except for the rear valance, the car has been primed. The black swirls are a scratch coat that reveals the low spots in the body work once the body is block sanded. Much to my pleasant surprise the rear quarter panels are complete and look great...

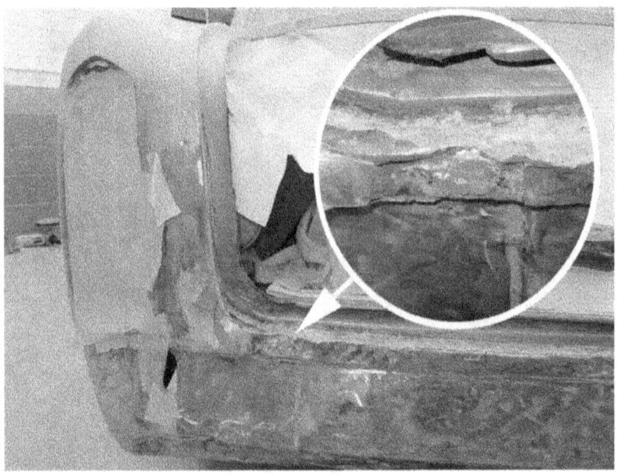

The rear valance is even uglier than originally thought. Crewe replacement parts are unavailable. There are several sources of rust repair panels, such as Healey Brothers in the UK, but these replacement panels do not extend high enough to cover the damage. The search is on for a salvaged part.

April 15, 2005

Doors, boot lid and bonnet have been mounted to check fit. I started the car today for the first time in several months. She fired up with some minor encouragement. The tappets quieted down rapidly with 40 psi oil pressure and the brake accumulators pressurized without undue hesitation...

Everything fits well except for the boot lid, which is a bit tight side to side. It closes nicely, but just happens to be cracked open in this photograph. I have finally located a proper rear valance at SC Works West in Newark, New Jersey and should have the part soon.

April 29, 2005

Scott has been busy block sanding and filling in the low spots...

The salvaged rear valance has arrived. Thanks to Rich at SC Works West for cutting the part out exactly as requested. I was slightly disappointed to see that it was a Shadow II with exhaust cutout and large bumper holes, but Scott says that he can work with this piece to fix the damage on my car. Also in the shipment: a set of left front fiberglass splash guards.

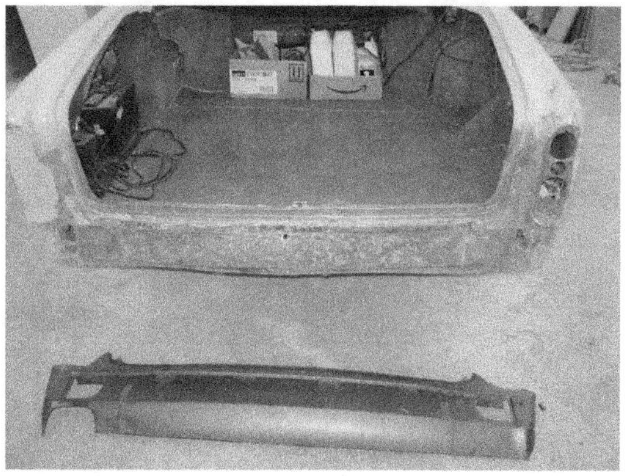

May 13, 2005

The rear valance area has been completely repaired...

Even the proper sized bumper holes have cleaned up nicely...

Front three-quarter view

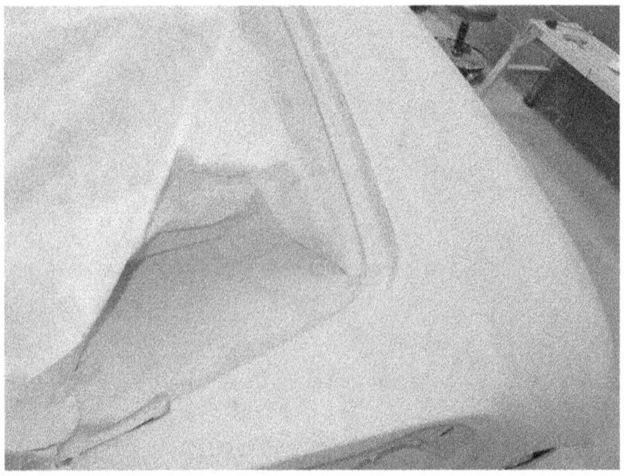

Even small areas in the bonnet channel are filled and sanded...

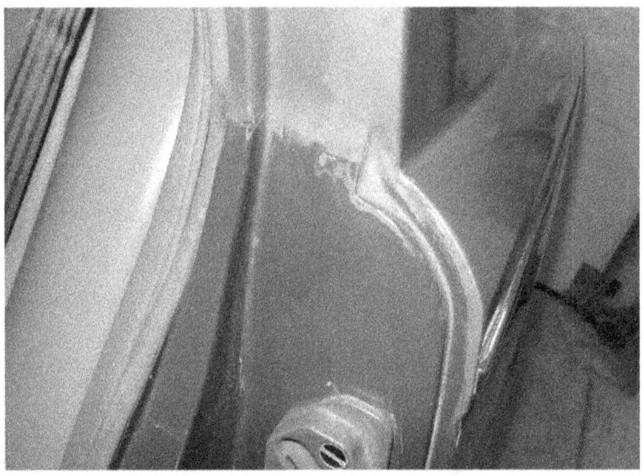

As it is getting close to painting time, the location of paint color transition lines becomes an important topic, particularly in the door jambs. Fortunately, the door jambs in my car were not resprayed along with the rest of the car in the Eighties.

May 20, 2005

Inboard sides of doors are also painted.

This photo reveals the Brewster Green color as it is in full sun. (Well, the *color* photo does, trust me!)

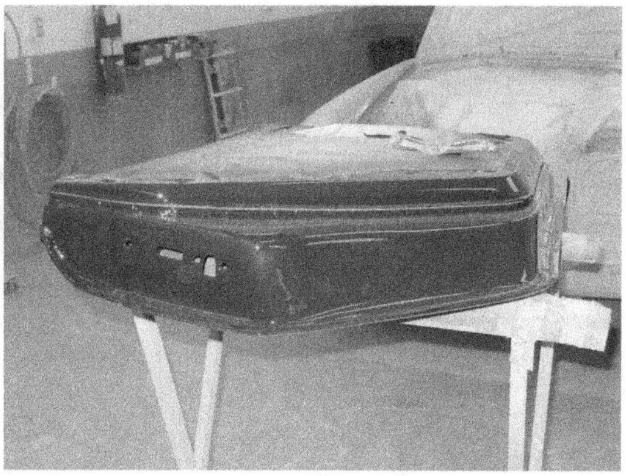

The car finally in paint: four coats of PPG Deltron DBC 2000 base coat followed by three coats of clear...

Wet sanding is next. Reassembly will start on the 31st.

June 2, 2005

Color transition looks great...

Trim starting to be reattached. How about that stainless steel polishing! Masking tape removed some of the leather finish in the door apertures. Looks like I'll be redying sooner rather than later.

June 7, 2005

Right side windows in. Lots of brightwork reattached.

June 9, 2005

Scott has driven "Elsinore" outside so that Paul can take a close look at the paint in daylight. I delivered the polished grille today. All that's missing is a pair of tail lamp seals, the rear bumper gaskets, new jam nuts for the wiper spindles and some plastic fasteners for the fresh air intakes. All but the fresh air items will arrive at my place by early next week...

Our only source of discontent at the moment is this weather strip which is about an inch too long between the top corners of the door. Oddly, the other rear door is fine. I'm waiting for Scott to install all four before I contact IntroCar Ltd. This won't hold up anything else, so work continues.

June 13, 2005

Scott's hiding behind the driver's door installing the last door seal. It fits, so it seems the ill fitting door seal issue is confined to only one door. The grille is on: an iconic moment. Not pictured: the sill strips are on and the boot lid carpet is glued in place. I can't wait to see the bumpers!

June 17, 2005

Front bumper is on and air intakes are in. Someone (Waples) got too excited and reversed the overriders left to right, so the bumper has to come off again to swap them.

Cowl and wood trim has been removed to ease windshield installation. It would be nice if the windshield guy showed up, though.

Hood pads have been installed. Sealing strips are in process. The elusive wiper spindle jam nuts (a.k.a. antenna nuts) have been installed.

June 24, 2005

We're even closer now. The rear glass is in. The wheel cover trim rings have been primed and are ready to paint. Today we decided that the coach lines will be ivory after Paul temporarily painted a stripe of that color on the body. Scott wants to move the right rear door forward 1/16" to 3/32" to improve the gap. Other things to do: weld up the driver's floor pan, put the new seal (cheerfully replaced by IntroCar Ltd.) on the left rear door and fix the horn (sounds funny)...

A bit of a tragedy on the windscreen: attempted installation resulted in breakage. Wednesday Albers thought they had a line on a new one already in the U.S. but when it arrived at their place they found it to be for a Silver Spirit or Spur. They ordered one from Crewe today and it should arrive in Ypsilanti via Zionsville, Indiana in seven days. Note that the front overriders are now properly oriented and all of the wheel well splash shields are in and blacked out. Scott fashioned a bracket for the electric antenna motor because he didn't like how it was "just hanging in there."

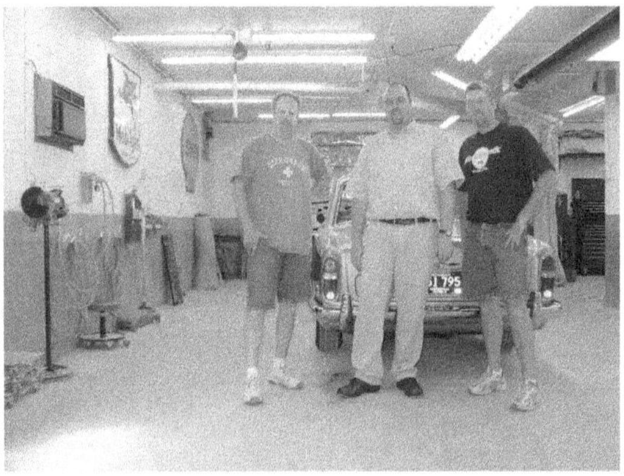

July 1, 2005

Today's the big day! The windshield is in and she's pinstriped in ivory. Here are Scott, Jon and Paul (l-r) just after a celebratory ale.

Chapter 7
Brakes and Hydraulics

Section 7-1
Fabricate Replacement Hydraulic Lines

The majority of the labyrinthine hydraulic pipe work is comprised of 3/16-in. Bundy tubing. The connections are either single flare (a.k.a. "Girling") or double flare. Both types can be made on flare tools that are designed for SAE 45 degree flares, which are the most common type of tool, at least in the U.S.

To fabricate a replacement brake or hydraulic line, you will need some 3/16-in. Bundy tubing, a 45-degree double-flaring tool, tube cutters, emery cloth, a set of small files and a workbench-mounted vise.

Cut the tubing to the right length using the tubing cutter. (See fig. 7-1-1.) Rotate the cutter around the tubing while steadily adding pressure by turning the knob. Bundy tubing is soft enough to gently bend into the desired shape by hand. You may choose to do some of this before forming the flares.

Preparing the tubing prior to the flaring operation is the key to success. Deburr the tubing end's inner and outer edges with a suitably shaped file or reamer. Make sure that the cut you made is 90 degrees to the tubing. If it is not, file it flat. Remove any burrs with emery cloth.

Insert the tubing into the correct size hole in the flaring bar. Make sure that the end to be flared is on the side of the bar with the countersunk holes. It must protrude above the top of the bar by a distance equal to the height of the step on the adapter, or in

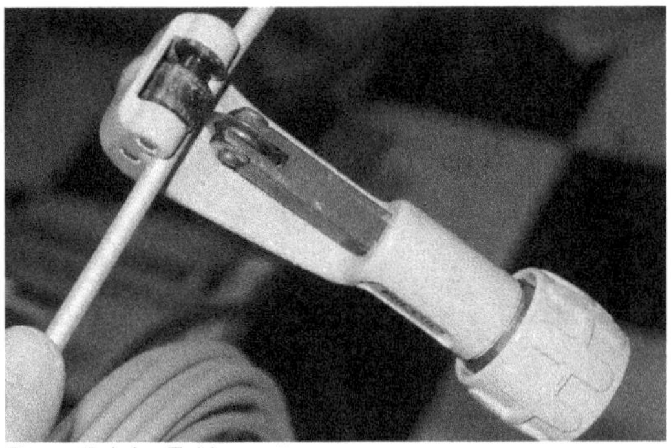

Figure 7-1-1 – Tubing cutter.

Figure 7-1-2 – Set proper tubing height.

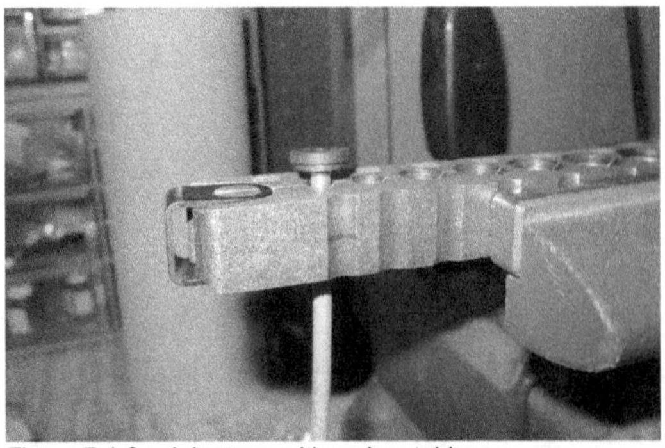

Figure 7-1-3 – Adapter positioned on tubing.

the case of the 3/16 adapter, the height of the first step. (See fig. 7-1-2.)

Tighten the screw that holds the flaring bar shut, which will, in turn, grip the tubing in place. Place the correct size adapter over the end of the tubing with the pin inserted into the inner diameter. (See fig. 7-1-3.)

Slide the yoke of the flaring tool over the flaring bar and carefully lower the cone until it comes into contact with the adapter. Slide the yoke along the flaring bar if necessary so that the cone lines up perfectly with the adapter. (See fig. 7-1-4.)

Screw the cone down until the adapter touches the flaring bar. You have now successfully made a single flare. (See fig. 7-1-5.) To make a double flare, release the cone far enough to remove the adapter. Screw the cone down again until it comes into tight contact with the tubing. The double flare is now complete. (See fig. 7-1-6.)

Assuming that you are flaring both ends of the tubing, slip the threaded fittings onto the tubing before forming the flare on the other end of the tubing.

Note: opening a hydraulic circuit introduces air into the system. It is important to bleed any circuits you've disturbed in order to restore correct operation after your work is complete. See "Bleeding the High Pressure System" on p. 35 in Section 2-4.

Figure 7-1-4 – Flare tool in position.

Figure 7-1-5 – Single flare.

Figure 7-1-6 – Double flare.

Section 7-2
Master Cylinder

As mentioned in Section 1-3, the purpose of the master cylinder is to provide brake feel and serve as a backup in the event of pressure loss in the two main braking circuits. The master cylinder gives little trouble during its useful life as long as the brake fluid is changed every two years. When it does expire after a quarter century or so, the seals fail, filling the protective boot with brake fluid.

If you have a Shadow with a chassis number earlier than 22118 that leaves brake fluid on the garage floor under the driver's seat, the master cylinder is the most likely culprit.

To check, remove the "rat trap" cover and it will be obvious if the seals have failed. The rubber boot on the master cylinder in fig. 7-2-1 has completely filled with brake fluid. Looking a bit closer in fig. 7-2-2, it is clear that the G-valve has also been leaking to some degree. Although the Workshop Manual explains in great detail how to overhaul the master cylinder, you can purchase a new unit from your Bentley dealer for about $80. You might consider rebuilding the G-valve at the same time, since it is easier to do so with the master cylinder out. Besides, the seal kit that includes a replacement valve seat, rubber seal and o-ring is only about $7.

It is important to pause and talk about safety. The inner workings of the "rat trap" are fundamental to brake operation and should not be treated lightly. This system is designed to stop a two and a half ton vehicle from 120 mph. Even at parking lot speeds, stopping the car is not something that you can do by turning the engine off and sticking your foot out the door. Once you start this job, the car is stuck where it is until completed. If you are not confident in attempting this job, find someone competent who is willing to help.

Figure 7-2-1 – Leaking master cylinder.

Figure 7-2-2 – Leaking G-valve.

Figure 7-2-3 – Homemade blanking plugs.

Ideally, you will have your car on a lift, or you can use the supplied jack on the driver's side of the car along with jack stands and wheel chocks for an added measure of safety. Position the jack stands with wooden blocks to support the trailing edge of the front arch and leading edge of the rear arch on the driver's side. You will need to disconnect a total of six hydraulic lines to perform this job: five male 3/8-24 and one male 7/16-20. You can make blanking plugs from commonly available fine pitch (UNF) hardware as shown in fig. 7-2-3. They may leak slightly and I would not recommend using them under pressure.

To reduce risk of injury or death while working underneath the vehicle, chock the wheels and disable the transmission by removing the gear shift cut out from the fuse board.

The master cylinder is at the rear of the "rat trap". (See fig. 7-2-4.) To remove it, loosen the two large retaining bolts and then disconnect the lines at the master cylinder. The rear line will flow a tremendous amount of fluid since it empties straight from the reservoir through a large ¼-in. Bundy tube—have the 7/16-in. diameter blanking plug at the ready! Once you have disconnected both of the master cylinder lines, bend them out of the way as little as possible to avoid alignment issues when reassembling.

Now you can remove the retaining bolts the rest of the way. Then carefully loosen the locknut on the actuating rod the

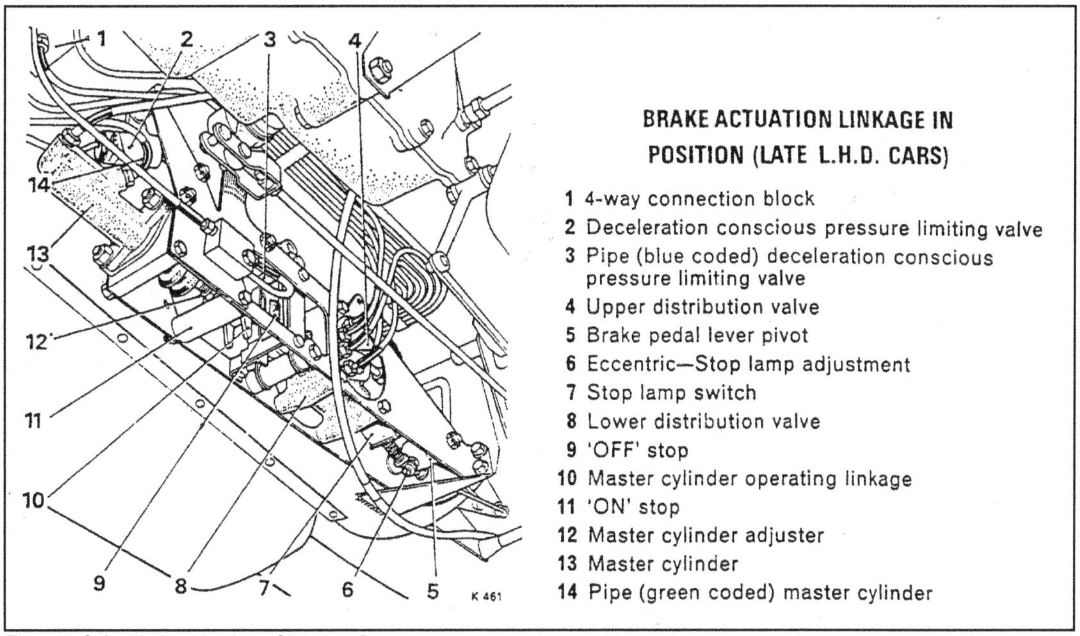

Figure 7-2-4 – Underside view of "rat trap."
(Illustration supplied courtesy of Bentley Motors, Inc.)

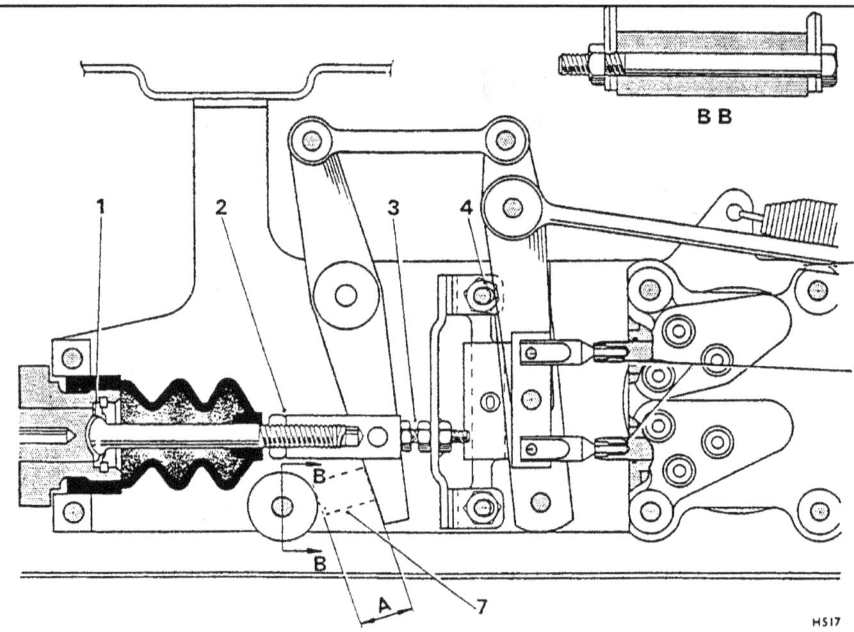

CURRENT ARRANGEMENT—ADJUSTABLE 'OFF' STOP—MASTER CYLINDER AND DISTRIBUTION VALVES WITH LARGER DIAMETER 'ON' STOP

1 0·005 in. to 0·010 in. (0,13 mm. to 0,25 mm.) clearance—push rod
2 Adjuster—'ON' stop setting
3 Adjuster—reducing master cylinder push rod free travel ('OFF' stop)
4 Elongated holes—adjusting distribution valve push rod travel
5 Zero clearance
6 Groove—upper face of valve
7 Gauge
A 0·800 in. minus 0·025 in. (2,032 cm. minus 0,64 mm.)

Figure 7-2-5 – Side view of brake actuation linkage in "rat trap."
(Illustration supplied courtesy of Bentley Motors, Inc.)

minimum amount needed. By doing this, you can use the locknut position as reference when the new unit is installed. Rotate the master cylinder and actuating rod to remove it from the "rat trap". Again, be careful not to disturb the locknut position.

Removal of the G-valve is covered in Section 7-3. Reassembly is a nervous affair of rapidly hemorrhaging brake fluid while you try to align and then tighten the brake lines. Now is not the time for one to forget how brake fluid will remove the paint in a flash—be careful! Ensure that all nuts and bolts are tightened; this is your braking system, after all. Replace all cotter pins with stainless steel and drop them in from the top or close to it as practicable. Trim the pins to minimize exposed sharp edges. Note that later Shadows may use locking tabs instead of cotter pins.

The Workshop Manual describes a very elaborate adjustment sequence to ensure that the brake distribution valves and master cylinder all actuate harmoniously. Among other things, it involves measuring the travel distance to the "on stop" using a gauge block and also ensuring a 0.005-0.010 in. gap on the master cylinder actuating rod when the brakes are not applied. (See fig. 7-2-5.)

If everything worked nicely prior to master cylinder failure, and you made sure that the adjustments in the rat trap were not disturbed in any way with the exception of the master cylinder actuating rod lock nut, you can avoid the lengthy Workshop Manual procedure.

Carefully measure the distance from the mounting face of the old master cylinder to the contacting face of the locknut. Transfer the lock nut to the new master cylinder and adjust it to the same distance. (See fig. 7-2-6.) Once the new master cylinder is installed with the same mounting face to locknut distance, it is only necessary to verify the 0.005-0.010 in.

If the G-valve or distribution valves were disturbed, bleed the high pressure hydraulic circuits per the Workshop Manual as reiterated in Section 2-4.

It is very difficult to bleed the master cylinder circuit by pumping the brake pedal. Using a Motive Products power bleeder or similar is much more effective. This type of bleeder pressurizes the brake reservoir to about 10 psi and forces the fluid through the master cylinder to an open bleed nipple. Ordinarily, the bleeder can be clamped in place on all manner of vehicles, but the Shadow reservoir is too large.

Rather than going though the trouble of modifying the clamping mechanism, it is easy enough to hold it in place with the tube positioned over reservoir #2 with the rubber gasket sealing the opening over reservoir #1. You will need to have a helper pump up the power bleeder and crack open each low pressure bleed nipple in turn. The two reservoirs share airspace so they both need to be sealed off. (See fig. 7-2-7.) Bleeding the low pressure circuit takes less than 10 seconds for each of the two bleed nipples on the rear calipers. The low pressure bleed nipple is the upper one on very early Shadows and the lower one on later

Figure 7-2-6 – Note locknut distance to mounting face.

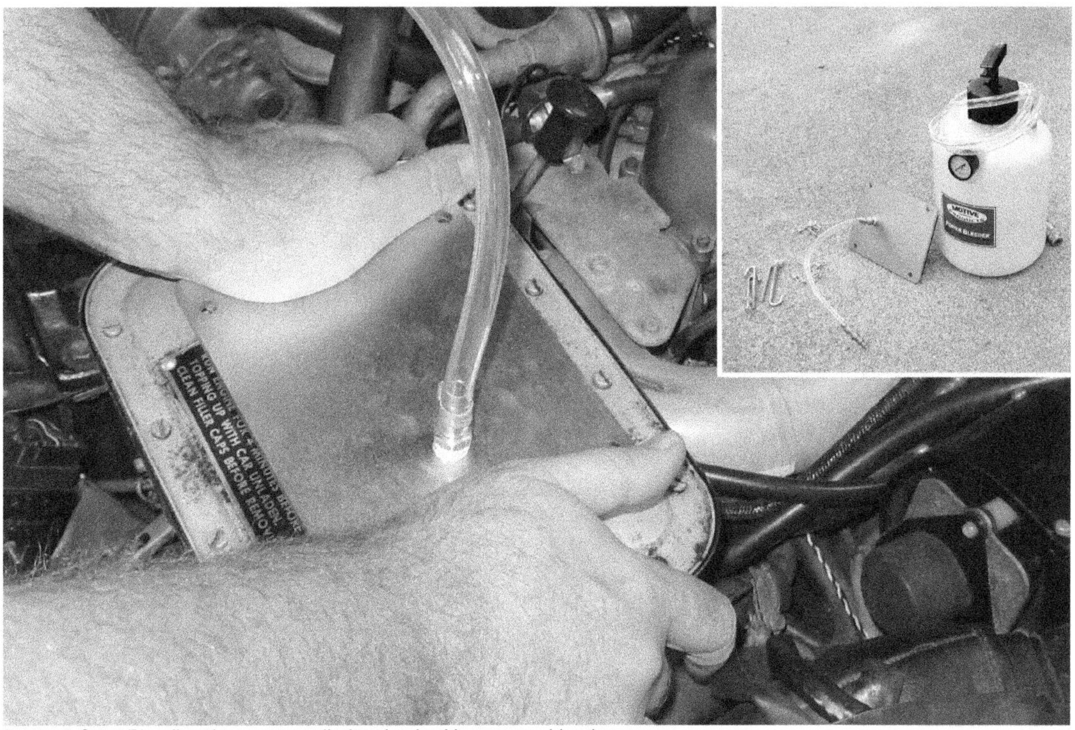

Figure 7-2-7 – Bleeding the master cylinder circuit with pressure bleeder.

Shadows from chassis number 1899 through 22117. (Subsequently, the master cylinder was dropped altogether. Refer to fig. 2-4-5 on p. 35.)

After starting the car to build pressure in all hydraulic circuits, it is a good idea to leave the "rat trap" cover off overnight to make sure there are no leaks. A minor drip from the distribution valves are acceptable, as they have no seal per se.

To learn more, I highly recommend Ralph Curzon's Silver Shadow Braking Seminar Video Series available from RROC Club Stores. The two part video series covers detailed refurbishment of the Shadow braking system in its entirety, with the exception of the master cylinder.

Section 7-3
Rebuild G-Valve

Officially known as the "deceleration conscious pressure limiting valve," the G-valve is found at the rear of the "rat trap" which is underneath the driver's seat on the underside of the vehicle. (See fig. 7-2-4.) It consists of a casting, a large steel ball on an incline, a small valve seat with rubber seal and two brake line connections. (See fig. 7-3-2.)

You will recall that there are two separate brake circuits feeding the upper and lower pistons on the rear wheels. The G-valve is in series with either the uppers or lowers depending on the year the car. The ball and the ramp are oriented in the "rat trap" so that strong braking will cause the steel ball to roll up the ramp and seal against the valve seat. When this happens, brake pressure is reduced to half of the rear braking system, preventing the rear wheels from locking and compromising directional stability.

The G-valve is located above the master cylinder on Silver Shadows so equipped. Consult Section 7-2 for master cylinder removal instructions. Disconnecting the two lines that pass closest to the G-valve at their nearby 4-way connection block makes removal of the G-valve easier. Removal of these lines will require four 3/8-24 diameter blanking plugs. It is easier to service later cars without a master. Remember that RR363 is an effective paint remover, so be

Figure 7-3-1 – "G" Valve removed from rat trap.

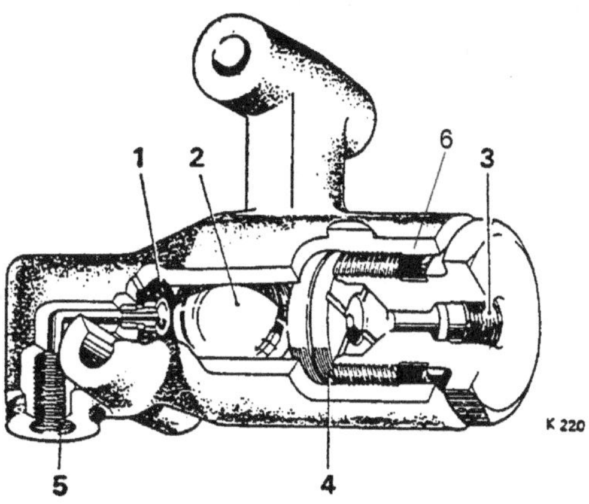

CUT-AWAY VIEW—PRESSURE LIMITING VALVE

1 Valve seat and rubber seal
2 Steel ball
3 Inlet port
4 Spacer plate
5 Outlet port (1-3/8 hex)
6 O-ring

Figure 7-3-2
(Illustration supplied courtesy of Bentley Motors, Inc.)

careful what you touch. Also take care not to introduce any contaminants into the braking system.

Once the G-valve is out, drain any excess fluid from it and clamp it in a vise so that the large hex head is accessible. Using a 1-3/8-in. socket and a large 1/2-in. drive ratchet, remove the outlet port, spacer plate and steel ball. Use a #2 EZ-out to remove the press fit valve seat and what remains of the rubber seal and discard them. Remove the o-ring from the outlet port and discard it. Clean everything that is left with denatured alcohol.

Place the new rubber seal on the new valve seat, with the seal oriented so that the molded-in writing will remain visible once the rubber seal and seat are tapped back into place. Tap the valve seat and seal into the casting with a small brass drift. Replace the steel ball and spacer plate. Install a new o-ring in the outlet port and tighten onto the casting.

Installation is the reverse of removal. Ensure that no contamination enters the brake system. Bleed the high pressure circuits as discussed in Section 2-4.

Section 7-4
Renew Leveling Valve Seals

There are two leveling valves on a Shadow rear suspension. Each one is attached to the car body adjacent to a rear trailing arm. Each trailing arm is connected to a link which rotates the operating shaft on the leveling valve. Each valve has four hydraulic connections: high pressure inlet, outlet to height control ram, reservoir return and signal pressure from the solenoid valve. The leveling valves are generally trouble free and only tend to leak after a decade or two of service.

Before discussing the procedure, a brief word on system operation is in order. The rear spring on each side of the car has a hydraulic ram mounted at the top that forces the spring downward as pressurized hydraulic fluid is introduced thereby raising the ride height of the vehicle. To control ride height, the leveling valve, in a sense, measures the distance from the trailing arm to the body. If the trailing arm is too close,

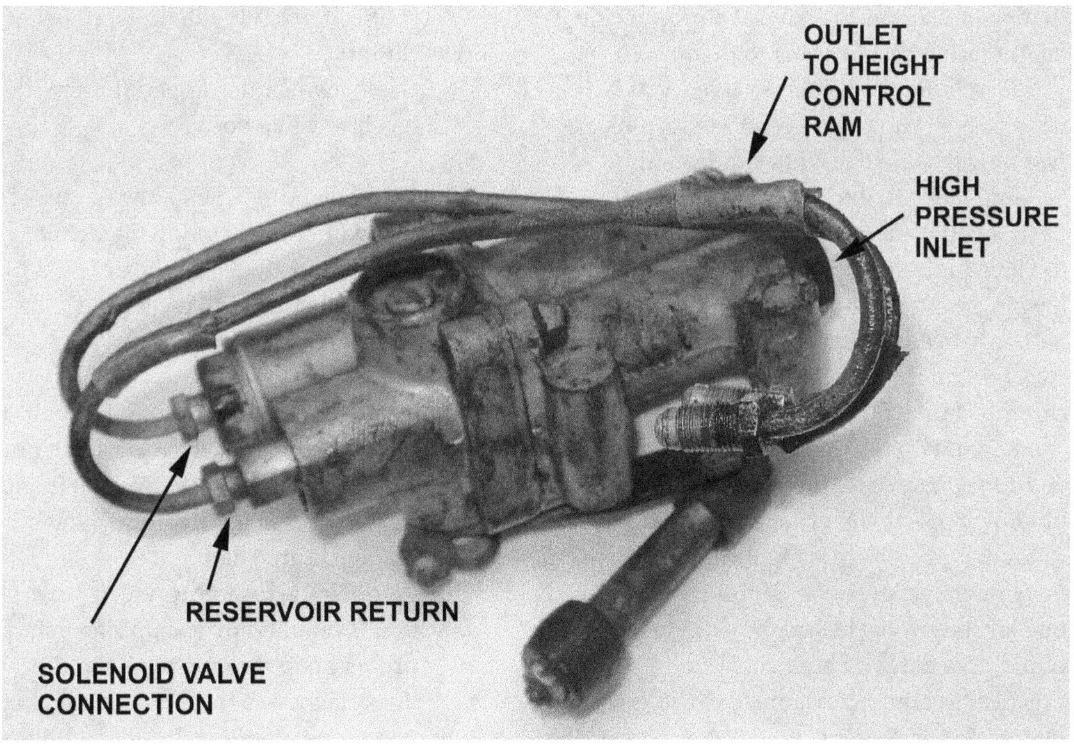

Figure 7-4-1 – Right hand leveling valve. Bodies of the left and right hand valves are identical except for the side of the body to which the linkage is attached. When mounted, the solenoid valve and reservoir return connections are toward the front.

the ride height is too low and additional fluid is passed to the height control ram. If the trailing arm is too far away from the body, the ride height is too high and the leveling valve allows some fluid from the ram back to the reservoir.

All of this works very nicely when the car is stationary, but when it is in motion, both trailing arms are in continuous up and down motion following every nuance of road surface. This is where the solenoid valve comes in. If the car is in park or neutral, or one of the doors is open, voltage is applied to the solenoid valve. This in turn supplies high pressure fluid to the leveling valve when the car is stationary and no pressure when underway.

The presence or absence of pressure at the solenoid valve connection causes a plunger to move inside the leveling valves, restricting the flow of fluid in and out of the ram. When the flow is restricted, the leveling function corrects ride height at a much slower rate, effectively *averaging* the road irregularities to arrive at a proper ride height rather than *reacting* to them.

To rebuild the leveling valves, you will need two leveling valve rebuild kits plus two new flexible lines that run from the leveling valve to the height control ram. Each flexible line can only be removed with the leveling valve already out. Since the leveling valve is a bit of a chore to remove and only needs to come out every decade or two, you might as well do both. Be careful—brake fluid removes paint.

Begin with the car on a four post lift. I would hesitate to do this job without this kind of access to the vehicle underside. The following procedure is identical for each leveling valve. Remove, rebuild and replace one leveling valve before moving on to the other. Use flare nut wrenches where practical to minimize damage to the hydraulic connections.

1. Ensuring that the system is depressurized as in Section 2-8, disconnect the front two leveling hard lines (solenoid connection and reservoir return) where they are connected at a junction block. (See fig. 7-4-1.)
2. Disconnect rear hard line at the leveling valve. (This is the high pressure inlet connection.)
3. Loosen the flex line at rear of valve. (This is the outlet to the height control.)
4. Disconnect the linkage at the lower end where it attaches to the trailing arm.
5. Remove the two nuts and bolts that attach the valve body to its mounting bracket.
6. To separate the leveling valve from the flex line, twist the leveling valve body while holding the flex line stationary with a flare nut wrench.
7. Remove the flex line from its chassis connection with a pair of wrenches. The bracket that supports this connection is very delicate, so be careful.
8. Drain away any trapped fluid in the leveling valve body.

Now is a splendid time to have a cup of coffee and a danish. The following detail work is best done with a clear head on a large *clean* workbench with a vast and uncluttered work surface.

9. Remove the linkage from the shaft, remembering which side it went on. The only difference between left and right valves is the side on which the linkage is attached.
10. Remove the reservoir return port that is retained by a snap ring using proper snap ring pliers. Keep the line attached to the end piece so that you do not confuse this line with the upper line.

Figure 7-4-2 – Removal of leveling valve piston.

11. Turn the shaft with screwdriver to release the piston. (See fig. 7-4-2.)
12. Remove the solenoid valve connection, ensuring that all of the washers are accounted for; these serve as shims of a very specific thickness for calibrated operation of the leveling valve.
13. Remove the high pressure inlet and outlet to height control ram fittings. They are not identical so ensure that you remember which is which. Make sure that you capture the springs and pistons contained inside the body behind each of these.
14. Remove the four nuts holding the two body halves together. Collect the piston, ring and shaft pieces.
15. Remove the spring clips on each side of the shaft.
16. Push the shaft in each direction to pop loose the dust cover.
17. With a pick, carefully remove the shaft seal.
18. Remove the two o-rings from the valve body halves and the o-ring from each of the four end pieces.
19. Carefully clean all of the parts in denatured alcohol and dry with compressed air.
20. Reassemble using new o-rings, smearing each with fresh RR363, using fig. 7-4-3 as a guide.
21. Reinstall in the car reversing the removal procedure.
22. Since the height control system recirculates brake fluid back to the reservoir, it is essentially "self bleeding" and should return to normal function after running the car for a brief time.
23. Once it has returned to normal function, adjust the attachment point of the linkage in the slotted hole on the trailing arm to achieve proper ride height.

For additional instruction on this and related subsystems, I recommend the two part series of videos on Shadow Hydraulics hosted by Ralph Curzon and distributed by RROC Club Stores.

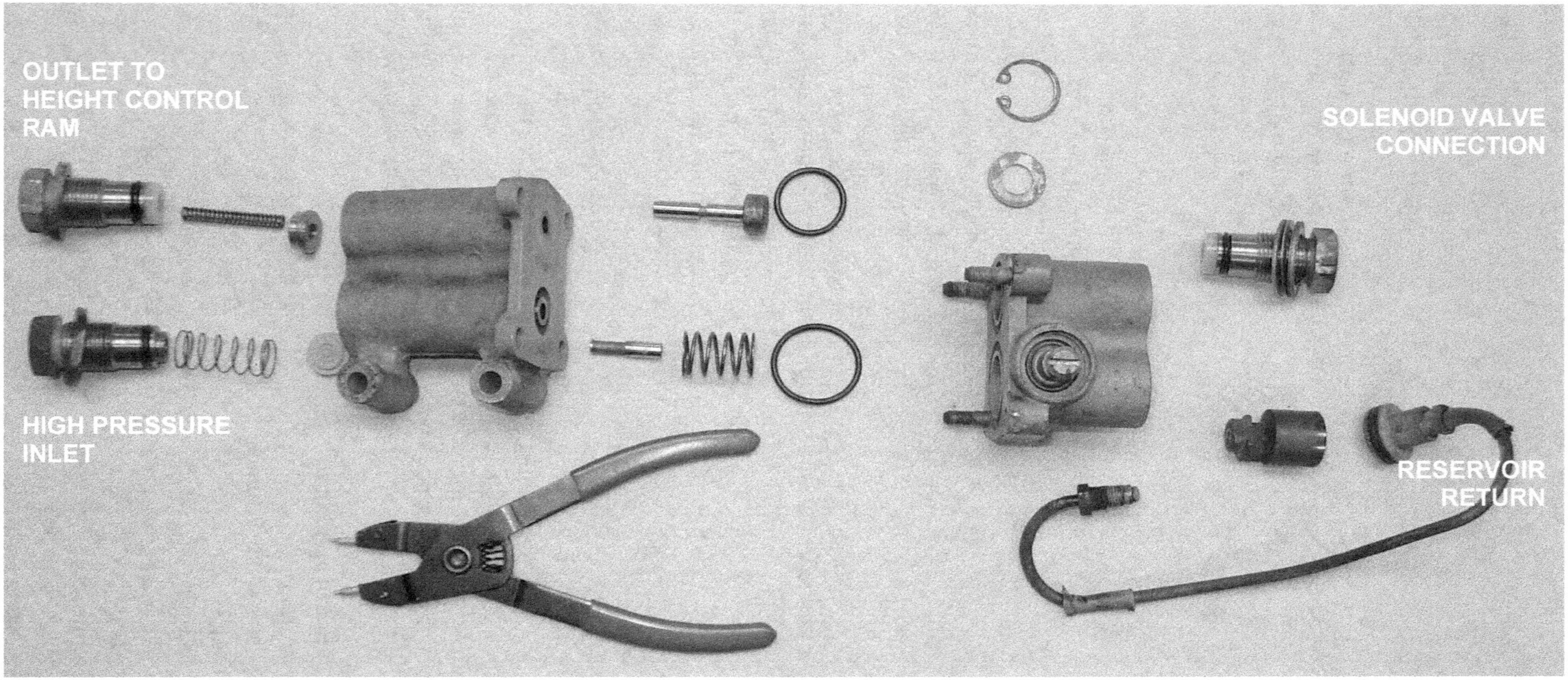

Figure 7-4-3 – Right hand leveling valve and its 30 plus parts.

Chapter 8
Climate Control

Section 8-1
Clean Duct Drains and Replace Leaf Screen

There are a series of drains responsible for draining water out of the cabin air handling system: two air intake trough drains, two dashboard duct drains and one A/C evaporator box drain.

Out of all of these, the dashboard duct drain tubes are by far the largest. Responsible for preventing water entry into the passenger cabin, they tend to clog with fine dirt. Water entering in this way is the primary cause for rusty floor pans. (See fig. 8-1-1.)

These L-shaped drain tubes are located in the engine compartment on the bulkhead on each side of the heater box. They are approximately the diameter of garden hose and are affixed to the bulkhead with a hose clamp.

Removal and cleaning of the right hand drain tube will be discussed here since the left side requires a subset of these steps. To perform this procedure on the left side, skip ahead two pages.

First, remove the electrical connections and washer tube from the top of the windshield washer bottle. Slide the washer bottle itself up and out of its bracket. (See figs. 8-1-2 and 8-1-3.)

Remove the two 2BA set screws, nuts and washers that hold the washer bottle bracket to the cruise control assembly.

Disconnect the cruise control chain at the throttle end. Make sure you mark which chain bead the throttle linkage was attached to. Remove the 7/16-in. A/F set screw and washer that holds the bottom of the

Figure 8-1-1 – Rust in footwell caused by blocked duct drains.

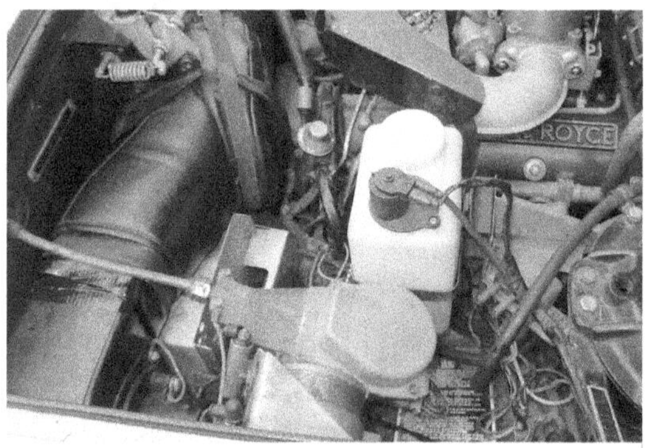

Figure 8-1-2 – Washer bottle prior to removal.

Figure 8-1-3– Washer bottle disconnected and removed.

Figure 8-1-4 – Lucas 16W wiper motor with covers removed.

wiper/cruise control bellows bracket to the fender apron.

Remove the two 1/2-in. A/F set screws and washer and one 1/2-in. A/F nut and washer retaining the wiper/cruise control bellows bracket to the fender apron.

Remove both electrical connectors from the windshield wiper motor. Remove the 2BA nuts from each of two condensers on the bracket and disconnect the wires. Be careful not to lose the fiber washers on the studs of the condensers. Note that the washers are used to prevent the wires from grounding to the case of their respective condenser.

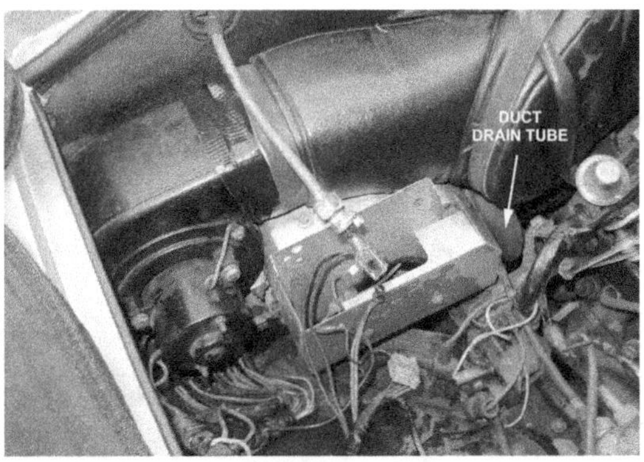

Fig. 8-1-5 – Drain tube becomes visible.

Remove the four sheet metal screws holding the plastic cover on the wiper motor.

Remove the remaining two sheet metal screws securing the metal cover on the wiper motor. Fully loosen the park position adjusting nut.

Lift the connecting rod so that it releases the cable. If the wiper motor is a Lucas 16W, carefully note the orientation of the slider block and remove it for safe keeping. (See fig. 8-1-4.) If the wiper motor is the earlier Lucas 6W which has a knurled parking switch adjuster adjacent to the cable, there is no slider block to worry about.

Remove the wiper/cruise control bellows unit in its entirety, disconnecting

Figure 8-1-6 – Blower motor housing, plenum and drain tube.

the vacuum line from the bellows on the way out. (See fig. 8-1-5.)

Note: if you are preparing to clean the left hand duct drain tube, begin here:

Remove the three 2BA nuts and washers securing the blower motor to its housing.

Disconnect the two motor wires. Remove the motor and squirrel cage. (See fig. 8-1-6.)

Remove the three 2BA nuts and washers securing the blower motor housing to the bulkhead and then remove the housing.

Disconnect the ground wire attached to the bonnet latch with a 7/16-in. A/F screw.

Release the strap holding the plastic duct to the heater box and then remove the duct.

Relax the hose clamp on the dashboard duct drain. Remove and clean. Probe the hole where the drain tube was attached with a finger. Clean as necessary. (See fig. 8-1-6.) Reassemble observing the following tips:

If you are addressing the right side duct drain, be sure to properly seal around the speedometer cable with dumdum.

Once the blower motor is reinstalled, check its function before proceeding with the remainder of the reassembly. Sometimes the squirrel cage will rub against the housing if not seated properly.

If you are addressing the passenger side duct drain, be sure to reinstall the fiber washers on the wiper motor condensers so they insulate the wire from the case. If you blow fuses when you test the wipers, look here first. There are two ways to reinstall the slider block on the wiper motor; the incorrect way will break the button on the proximity switch. The correct orientation is to locate the ramp on the slider block facing down and to the right as shown in fig. 8-1-4.

Finally, replace the urethane foam filter under the air intake grille at the base of the windshield. This will help to keep your duct drains clean. To replace it, remove the air intake grille which is held in place with five 2BA screws in the "cat walk" along the front.

Section 8-2
Diagnosing Blower Motor Issues

The blower motor circuit on Series I Silver Shadows consists of a center console fan switch, toe board connectors on the left side of the engine compartment, three large resistors mounted to the left side fender apron near the headlamp and two blower motors with noise suppression capacitors mounted to the bulkhead. (See fig. 8-2-1.) For chassis numbers between 9001 and 26708 inclusive there is a blower motor relay as well.

The blower motor circuit in the automatic system found on the Silver Shadow II is not covered in this section. It is transistorized and does not share any common elements with the earlier system other than the blower motors themselves.

<u>*Both Blower Motors Inoperative*</u>

If the blower motors are inoperative at all speeds, verify that the ACU switches are set to heat or cool. If so, check the fuse

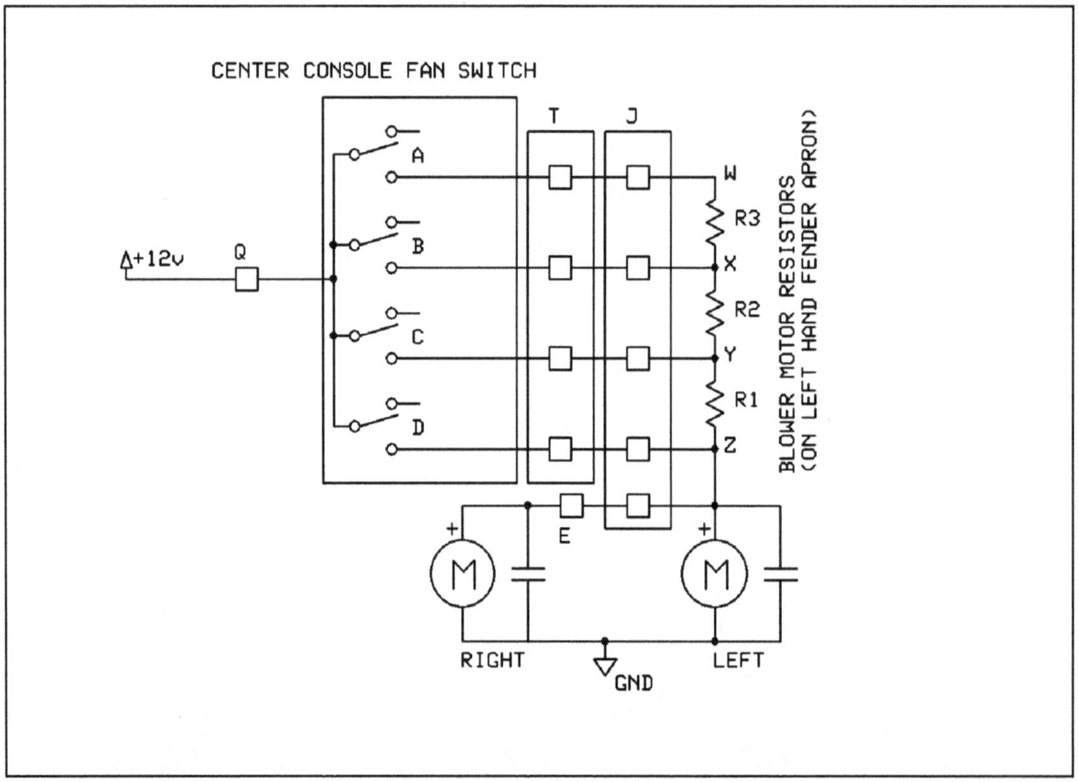

Figure 8-2-1 – Shadow I Blower Motor Circuit Schematic. Notation is consistent with Rolls-Royce functional schematic.

FAN SPEED	MICROSWITCHES A	B	C	D	TEST POINTS (Volts) W	X	Y	Z	CURRENT (Amps)
OFF	OFF	OFF	OFF	OFF	0.0	0.0	0.0	0.0	0.0
1	ON	OFF	OFF	OFF	11.7	9.4	7.7	6.0	0.8
2	ON	ON	OFF	OFF	11.7	11.7	9.6	7.4	1.0
3	ON	ON	ON	OFF	11.7	11.7	11.7	9.1	1.2
4	ON	ON	ON	ON	11.7	11.7	11.7	11.7	1.5
		Voltage	*11.7*	*Vdc*		*R2*	*2.2*	*Ohms*	
		R1	*2.2*	*Ohms*		*R3*	*3*	*Ohms*	

Figure 8-2-2 – Expected test point voltages at each fan speed setting.

board for a blown fuse. For chassis numbers 9001 through 26708, also check that the blower motor relay is functioning properly. If the relay checks out, suspect the Air Conditioning Control Unit (ACU) which "authorizes" the relay to close only when heating or cooling. (The ACU is discussed in detail in Section 8-4.)

<u>*Blower Motor Circuit Operation*</u>

If the blower motors operate at some speeds, the issue will likely lie with the console fan switch, toe board connector, or blower motor resistors. (See fig. 8-2-1.)

The center console fan switch consists of a cam mechanism that progressively actuates four microswitches as higher fan speeds are selected. As shown in the left half of fig. 8-2-2, when the console fan switch is turned to the "OFF" position, all microswitches ("A" through "D") are open. Moving the center console fan switch one click down to position "1" closes microswitch "A". Moving the center console fan switch another click down to position "2" closes microswitch "B" in addition to the "A", and so on. Selecting position "4", which is maximum fan speed, causes all four microswitches to be closed.

Referring to fig. 8-2-1, when position "1" is selected on the center console fan switch, current passes from the blower motor relay (if so equipped), through center console connector "Q", through microswitch "A", through center console connector "T", through the left hand toe board connector "J" and through R3, R2 and R1 in succession and then splits to the left hand blower motor directly and the right hand blower motors via toe board connectors J and E.* Since the motors are rubber isolated, each one is grounded to the chassis via a wire in their respective wire harness.

When position "2" is selected on the center console fan switch, current passes from the blower motor relay (if so equipped), through the center console connector "Q", through microswitch "B", through center console connector "T", through the left hand toe board connector "J" and through "R2" and "R1" shown in fig. 8-2-1, and then to both blower motors, etc. (Even though it is closed, current does not pass through "R3" because the *voltage difference* between connections "W" and "X" is zero and thus, no current flows.) Since the resistance in the blower motor circuit is R2 + R1 instead of R3 + R2 + R1 as in the previous example, the motors turn faster.

When position "4" is selected on the center console fan switch, current passes from the blower motor relay (if so equipped), through center console connector "Q", through microswitch "D", through center console connector "T", through the left hand toe board connector "J" and then to both blower motors, etc. All of the

* The LUCAR connector blocks identified are correct for the majority of the the Series I Shadows. Consult the functional wiring diagram for your specific model to be sure.

resistors are bypassed in this mode and the blower motors receive full battery voltage.

Blower Motor Circuit Diagnosis, Jumper "W" through "Z"

A good first step in diagnosing Shadow I blower motor faults is to quickly eliminate the half of the system that is unlikely to be at fault.

With the engine off, attach a jumper with alligator clips at each end to the positive red connector on the alternator. *Be very careful not to touch the chassis with the other end of the jumper as this connection will carry enough current to burn your hand and possibly start a fire.* Touch each of the terminals of the blower motor resistors mounted on the fender apron in order to ascertain how they are interconnected. Fig. 8-2-4 shows a representative resistor arrangement, yours may be slightly different.

Assuming the blower motors spring to life when you do this, the terminal that

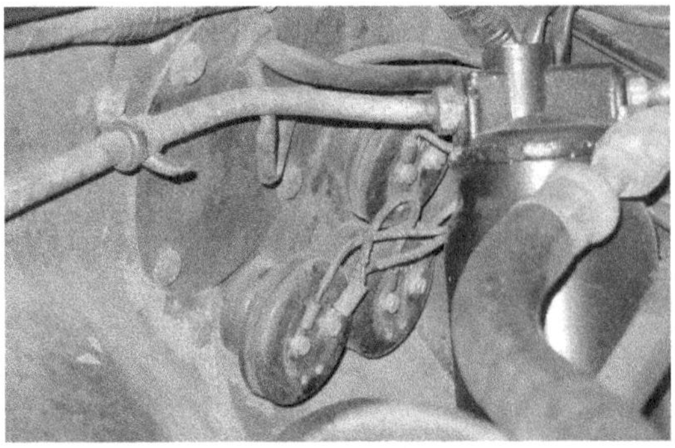

Figure 8-2-3 – Blower motor resistors on fender apron.

yields the slowest motor speed is point "W" and the next slowest is point "X" and so on. Each blower motor has two connections: positive and ground. The positive terminal for each motor has two wires because it is also connected to a noise suppression capacitor. The ground wire on each motor is a single wire. Jumpering either positive motor connection to battery voltage should also yield the fastest motor speed.

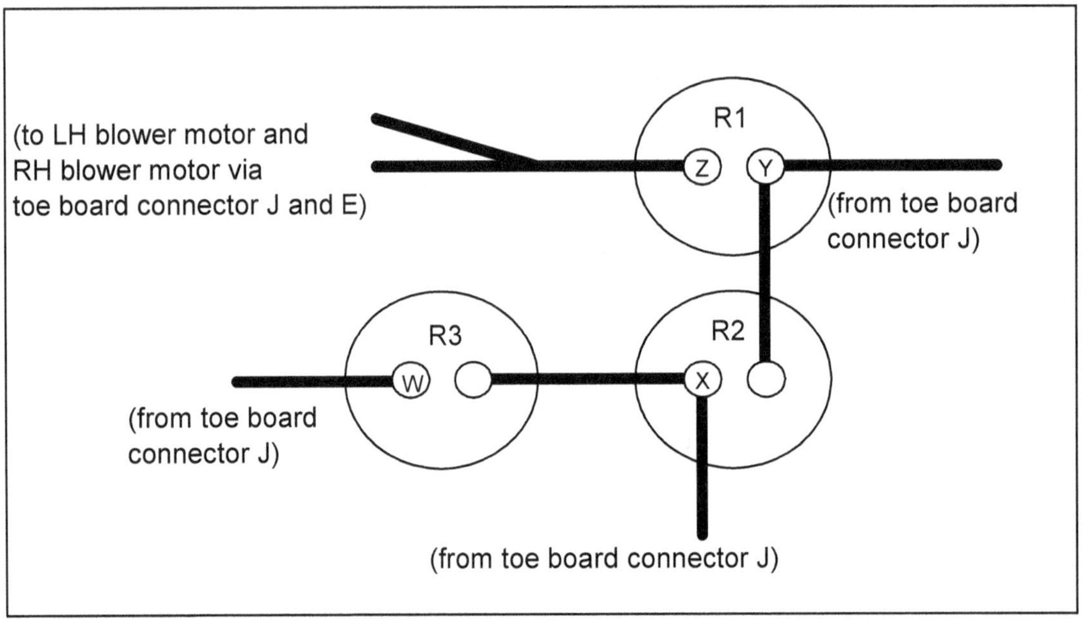

Figure 8-2-4 – Representative resistor arrangement on left hand fender apron.

"W" through "Z" and both Positive Motor Terminals Check OK

The most likely suspect in this case is an issue with left hand toe board connector J.

To gain working access to the bank of left hand toe board connectors, the left hand blower motor housing needs to be removed first. (See Section 8-1.) If, for example, the lowest speed is inoperative, find the color of the wire at point "W" on the resistor network and then find the same color of wire at toe board connector J. For chassis numbers 6000 and higher, you will find the Connector J contains three wires headed to the resistors and the return wire that goes to E for the right hand blower. Remove the connector, inspect both sides for damage and clean the male and female elements. While the connector is disconnected, measure the resistance of the affected wire from the connector to point "W" on the resistor network. The resistance should be 0.1 Ω or less. (See fig. 8-2-4.)

Because they are well protected from the elements, the center console fan switch and connections in the cabin are not often found to be the cause of blower motor malfunction. If you feel you have exhausted all of the other possibilities, remove the center console wood trim as in Section 5-3 and remove the center console side panels by prying as shown in fig. 8-2-6 using a screwdriver wrapped in tape to protect the covering. From this point, the blower motor switch can be disconnected from the console harness and removed. Use the resistance function of your multimeter to check the proper functioning of the switch as well as the continuity of each switch wire to the toe board connector.

"W" through "Z" not OK, both Positive Motor Terminals Check OK

In this case, there is a broken resistor or connection between resistors. Referring to the jumper test performed above, consult fig. 8-2-7. Use the "cause" column as a guide for what piece of the blower motor resistor network to remove and bench test. Approximate resistance values are shown in 8-2-2.

"W" through "Z" and both Positive Motor Terminals not OK "E"

If both motors do not spring to life when they are jumpered directly with battery power, the culprit is two broken motors or two broken grounds. To verify that the ground wire to each blower motor is

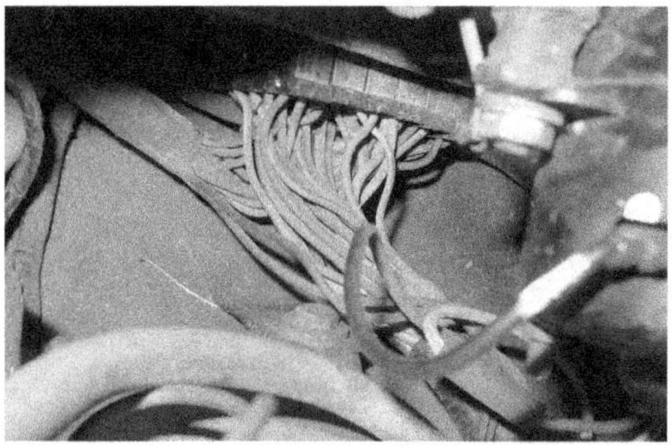

Figure 8-2-5 – Left hand toe board connectors.

Figure 8-2-6 – Removing center console side panels.

Test Result	Cause
Motors Inoperative "W" through "Y", OK at "Z"	R1 broken.
Motors Inoperative at "W" and "X", OK at "Y" and "Z"	R2 or wire between R2 and R1 broken.
Motors Inoperative at "W", OK at "X" through "Z"	R3 or wire between R3 and R2 broken.

Figure 8-2-7 – Resistor network diagnosis.

truly grounded, disconnect the ground wire from the blower motor. The ground wire is on the side of the motor that is not also connected to a noise suppression capacitor. Measure the resistance from the open end of this wire to a solid chassis ground such as the hood latch. It should read less than 0.1 Ω. A poor ground will reduce the available voltage at all speed settings. The resultant voltage at the motor may be insufficient to turn the motor at the slower speed settings. Correct the cause of the poor ground and reconnect the ground wire.

Section 8-3
Replace Heater Tap

In the course of getting my air conditioning online prior to the 2005 National Meet, I discovered that the system would blow extremely cold air for the first few minutes in the morning and then dramatically lose its effectiveness. Finally it occurred to me that the air conditioning was dying at the same time the engine thermostat was opening. It seemed that the heater tap was allowing hot water into the heater matrix regardless of the temperature setting. Perhaps this has happened to you?

On Shadow I and Shadow II vehicles up to chassis number 35000 or so, the heater tap is located on or near the rear support of the brake fluid reservoir. (See fig. 8-3-1 for Shadow I configuration.) On later Shadow II's, the tap is located near the lower radiator hose on the engine.

To confirm failure of the heater tap, turn the air conditioning system on by turning the upper temperature control knob fully counterclockwise. Beginning with a cold engine, start it and wait for it to come to operating temperature and the engine thermostat to open. You can confirm that the thermostat has opened when the top part of the radiator becomes too hot to touch.

Now feel the hose that exits the firewall side of the water tap. If it is also too hot to touch, the heater tap is somehow allowing hot water to enter the heater matrix. To confirm that it is the tap itself and not the servo actuator, turn the air conditioning off and the heat on and observe the servo moving the heater tap. If the servo moves the lever correctly, then it is the heater tap that needs to be replaced.

To replace the heater tap on your Shadow I, remove the pinch bolt that secures the small bell crank to the shaft of the servo actuator and then slide it from the shaft. Next, loosen and disconnect the firewall side hose from the heater tap. Be prepared to make a small mess with coolant leaking, but since it is at the top of the system, very little will be lost.

Figure 8-3-1 – Old heater tap in "off" position attached to rear leg of brake fluid reservoir. Lucas servo actuator is in the foreground at the bottom of the photo.

Remove the two 3/8-in. A/F nuts, bolts and washers securing the tap housing to the brake reservoir support. Be careful not to dislodge the copper tubes leading from the expansion valve to the air conditioning return pipe and suction throttling valve. Disconnect the remaining heater hose from the heater tap and remove the latter along with the associated cable and lever.

Remove the cable and lever from the heater tap. You may need to pry the tap lever open slightly to get the barrel shaped cable termination to come free. The remaining assembly consists of the heater tap itself, an o-ring, and an adapter housing. The adapter housing needs to be removed from the old tap and reused. In order to twist them apart, you may need to place the assembly in a vise and heat it with a torch. In the case of fig. 8-3-2, part of the tap mechanism remained lodged in the adapter. If this occurs, use a hammer and brass drift to push it out.

Clean the adapter with a wire brush so that it will present a clean surface to the new o-ring. File the mounting face of the adapter and the new heater tap so they will mount flush. (See fig. 8-3-3.) Sandwich the new heater tap, o-ring, and adapter together. Attach the cable to the tap lever. Again, slight deflection of the lever might be necessary.

Place the heater tap in position and connect the hoses. If the new heater tap is a reproduction part, the body casting differs slightly from the original. Because of this, it will be necessary to insert the two 3/8-in. A/F mounting bolts from the rear (rather than the front) in order to avoid interference. Installing the top nut and washer is straightforward; however, the bottom nut and washer required an inordinate amount of time and patience to get threaded on properly. Use a blob of wheel grease on the tip of a finger to hold the nut while reaching in around all of the obstructions. Childrens' hands would be a bonus in this situation.

Figure 8-3-2 – Old tap has disintegrated internally.

Figure 8-3-3 – Tap and adapter ready for installation.

Once everything is in place, push the lever back onto the servo. With the key in the "on" position but with the engine off, turn the left hand temperature control knob fully counterclockwise. This will ensure that the servo is commanded to the "off" position. Adjust the lever so that the heater tap lever is pulled fully outboard. (Refer to Figure 8-3-1.) Install and tighten the pinch bolt and replace any lost coolant. Start the engine and check for leaks.

Section 8-4
Air Conditioning Control Unit and Flaps

Prior to the Newport Meet in 2003, I asked shop owner and RROC member Dan Docherty to recharge my Shadow air conditioning system for the high speed trek across lower Ontario, New York and Massachusetts. He gently suggested that a lack of refrigerant was likely the least of my worries and that I should first ensure that all of the flaps and actuators were behaving as they should.

This section assumes that you have previous experience in troubleshooting electrical circuits. For a refresher, see Section 9-5.

The Shadow air handling system, as with most modern vehicles, consists of a heater matrix, A/C evaporator, heater tap (or "coolant valve") and ductwork with flaps designed to direct the quantity and temperature of air desired. (See fig. 8-4-1.) Rather than vacuum actuation of the flaps, suction throttling valve and heater tap, the Rolls-Royce system uses nine Lucas electric

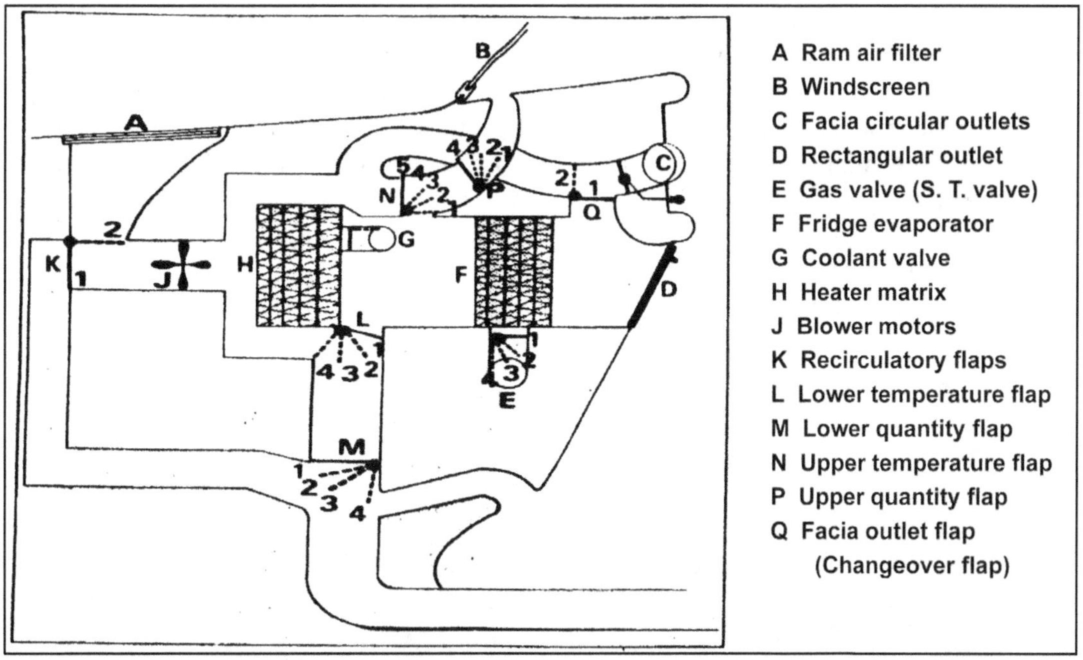

Figure 8-4-1 – Air handling scheme in Silver Shadow and Bentley T 1966-1976
(Illustration supplied courtesy of Bentley Motors, Inc.)

servo actuators controlled through the Air Conditioning Control Unit (ACU) in the center console.

Whereas the Silver Shadow I ACU consists of a system of mechanical contacts, the Shadow II ACU is fully electronic and is not within the scope of this book.

Air Conditioning Control Unit

The ACU consists of two rotary switches with multiple stacks to address the rather complex logic associated with temperature selection. In addition, there is a multifaceted conductive "pineapple" located at the end of each shaft for quantity selection. (See fig. 8-4-2.)

Consulting fig. 8-4-3, one can appreciate the complexity of the ACU. For example, assume the desired function is maximum defrost. This would be accomplished by twisting the upper quantity knob to position "4" and pulling it to position "3". In this case, the heater tap will be open (to position "2"), the upper quantity and temperature flaps will be fully open (to positions "4" and "5", respectively), and the two recirculation flaps will be open (to position "2"). Moving the upper knob four clicks counter-clockwise to the minimum A/C position keeps the upper quantity flap fully open, engages the A/C compressor and closes the recirculation flaps (to position "1") and coolant tap (to position "1") and switches over the circular outlets to receive refrigerated air (to position "2"). Turning the upper knob further counter-clockwise overrides the quantity setting by closing the upper quantity flap (from position 4 to position 1) to prevent windshield fogging.

Verifying the logic of the ACU is painstakingly slow. Figure 8-4-4 provides a table of pin outs for each of the two ACU connectors. This allows the function of the ACU and the rest of the system be tested independently.

Electric Actuators

The nine electric servo actuators are identical components. (See fig. 8-4-5.) Inside each unit, there are a series of concentric conductive tracks that rotate with the output shaft. Each track has a gap in it corresponding to a different stopping position. When power is applied to one of the pins on the connector, power flows to a finger sliding on the corresponding track which energizes the motor and causes the track to rotate until the finger reaches the gap and the connection is broken.

To test an actuator on the bench, ground the pin marked "E" and apply 12V power to the each of the pins, one at a time, ensuring that the output shaft rotates and then comes to a stop in each position. If the rubber boot is removed, you can also run the motor continuously by applying power to the red wire.

Diodes

A complete electrical diagram of the

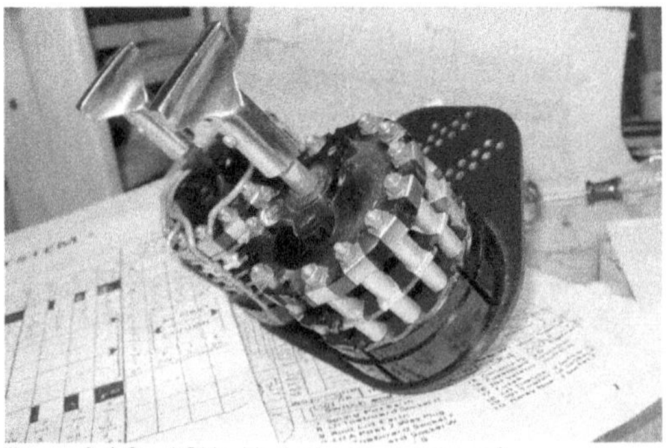

Figure 8-4-2 – ACU with dust covers removed.

Figure 8-4-5 – Lucas actuator.

		FRESH AIR				HEATED AIR																CONDITIONED AIR						
SWITCH	Twist	0	0	0	0	1	1	1	1	2	2	2	2	3	3	3	3	4	4	4	4	-1	-1	-1	-1	-2	-3	-4
POSITION	Pull	0	1	2	3	0	1	2	3	0	1	2	3	0	1	2	3	0	1	2	3	0	1	2	3	any	any	any
UPPER SYSTEM ACTUATOR*	Upper Qty.	1 UW	2 UG	3 UR	4 UB	1 UW	2 UG	3 UR	4 UB	1 UW	2 UG	3 UR	4 UB	1 UW	2 UG	3 UR	4 UB	1 UW	2 UG	3 UR	4 UB	1 UW	2 UG	3 UR	4 UB	1 UW	1 UW	1 UW
	Upper Temp.	1 YU	1 YU	1 YU	1 YU	2 YW	2 YW	2 YW	2 YW	3 YR	3 YR	3 YR	3 YR	4 YN	4 YN	4 YN	4 YN	5 YB	5 YB	5 YB	5 YB	1 YU	1 YU	1 YU	1 YU	1 YU	1 YU	1 YU
	Heater Tap	1 WY	1 WY	1 WY	1 WY	2 WP	2 WP	2 WP	2 WP	2 WP	2 WP	2 WP	2 WP	2 WP	2 WP	2 WP	2 WP	2 WP	2 WP	2 WP	2 WP	1 WY	1 WY	1 WY	1 WY	1 WY	1 WY	1 WY
	Circular Outlet	1 PY	1 PY	1 PY	1 PY	1 PY	1 PY	1 PY	1 PY	1 PY	1 PY	1 PY	1 PY	1 PY	1 PY	1 PY	1 PY	1 PY	1 PY	1 PY	1 PY	2 PU	2 PU	2 PU	2 PU	2 PU	2 PU	2 PU
	Recirc. Flap	1 RB	1 RB	1 RB	1 RB	1 RB	1 RB	1 RB	1 RB	1 RB	1 RB	1 RB	1 RB	1 RB	1 RB	1 RB	1 RB	1 RB	1 RB	1 RB	1 RB	2 RU	2 RU	2 RU	2 RU	2 RU	2 RU	2 RU
	Recirc. Flap	1 RG	1 RG	1 RG	1 RG	1 RG	1 RG	1 RG	1 RG	1 RG	1 RG	1 RG	1 RG	1 RG	1 RG	1 RG	1 RG	1 RG	1 RG	1 RG	1 RG	2 RY	2 RY	2 RY	2 RY	2 RY	2 RY	2 RY
	ST Valve	1 BR	1 BR	1 BR	1 BR	1 BR	1 BR	1 BR	1 BR	1 BR	1 BR	1 BR	1 BR	1 BR	1 BR	1 BR	1 BR	1 BR	1 BR	1 BR	1 BR	1 BR	1 BR	1 BR	1 BR	2 BW	3 BU	4 BN
	Air Volume	•	••	•••	••••	•	••	•••	••••	•	••	•••	••••	•	••	•••	••••	•	••	•••	••••	•	••	•••	••••	•	•	•
	Air Temp.	○	○	○	○	•	•	•	•	••	••	••	••	•••	•••	•••	•••	••••	••••	••••	••••	❄	❄	❄	❄	❄❄	❄❄❄	❄❄❄❄
LOWER SYSTEM ACTUATOR*	Lower Qty.	1 PW	2 PR	3 PN	4 PB	1 PW	2 PR	3 PN	4 PB	1 PW	2 PR	3 PN	4 PB	1 PW	2 PR	3 PN	4 PB	1 PW	2 PR	3 PN	4 PB							
	Lower Temp.	1 NY	1 NY	1 NY	1 NY	2 NG	2 NG	2 NG	2 NG	3 NU	3 NU	3 NU	3 NU	4 NR	4 NR	4 NR	4 NR	5 NB	5 NB	5 NB	5 NB							
	Heater Tap	1 WY	1 WY	1 WY	1 WY	2 WP	2 WP	2 WP	2 WP	2 WP	2 WP	2 WP	2 WP	2 WP	2 WP	2 WP	2 WP	2 WP	2 WP	2 WP	2 WP							
	Air Volume	•	••	•••	••••	•	••	•••	••••	•	••	•••	••••	•	••	•••	••••	•	••	•••	••••							
	Air Temp.	○	○	○	○	•	•	•	•	••	••	••	••	•••	•••	•••	•••	••••	••••	••••	••••							

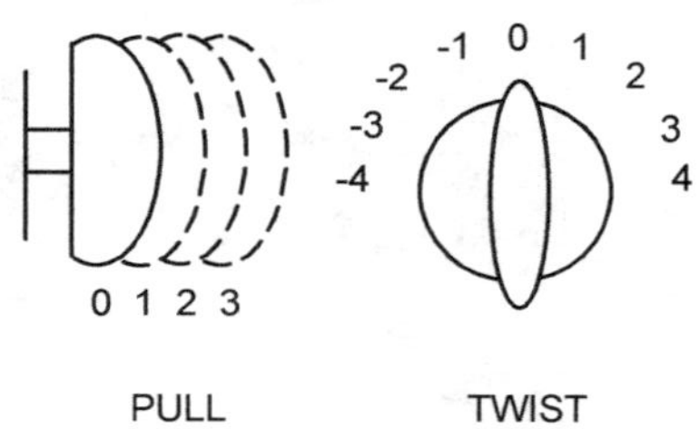

COLOR CODES FOR WIRING: B=BLACK, G=GREEN, N=BROWN, P=PURPLE, R=RED, U=BLUE, W=WHITE, Y=YELLOW
*NUMBER ABOVE COLOR CODE INDICATES WHICH PIN IS "HOT" ON THE ACTUATOR

Fig. 8-4-3 – ACU Logic Diagram

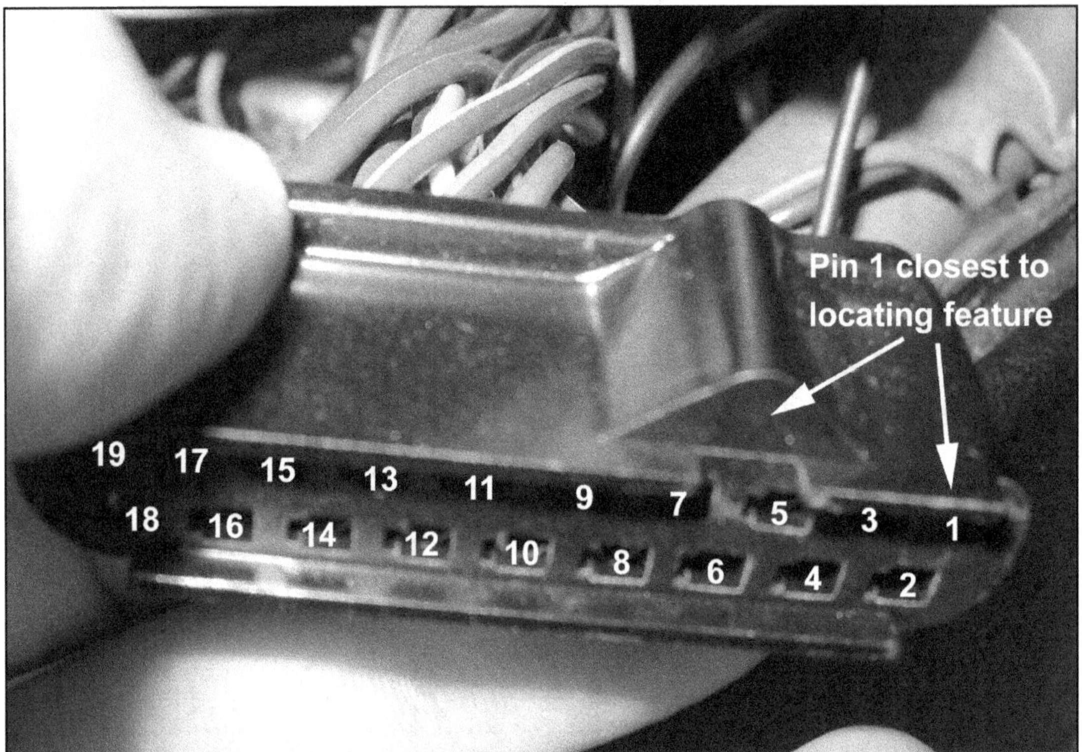

Left Hand ACU Connector (shown above)

1 Gas valve 4 (BN)
2 Upper temp. 1 (YW)
3 Upper qty. override to 1 (UW)
4 Upper qty. 1 (UN)
5 Circular flaps 2 (PU)
6 Upper temp. 0 (YU)
7 Coolant tap 2 (WP)
8 Upper qty. 3 (UR)
10 Gas valve 1 (BR)
11 Upper qty. 2 (UG)
12 Upper temp. 3 (YN)
13&14 Recirc. flaps 1 (RB/RG)
15 Circular flaps 1 (PY)
16 Gas valve 2 (BW)
17 Upper qty. 4 (UB)
18&19 Recirc. flaps 2 (RU/RY)

Right Hand ACU Connector (similar)

1 Lower temp. 1 (NG)
2 Lower temp. 2 (NU)
3 Lower qty. 1 (PW)
4 Lower temp. 3 (NR)
5 Lower temp. 0 (NY)
6 Lower temp. 4 (NB)
7 Lower qty. 3 (PN)
8 Gas valve 3 (BU)
10 Upper temp. 2 (YR)
11 Lower qty. 4 (PB)
12 Upper temp. 4 (YB)
13 Lower qty. 2 (PR)

Figure 8-4-4 – ACU connector pinouts. (See 8-4-3 for color code info.)

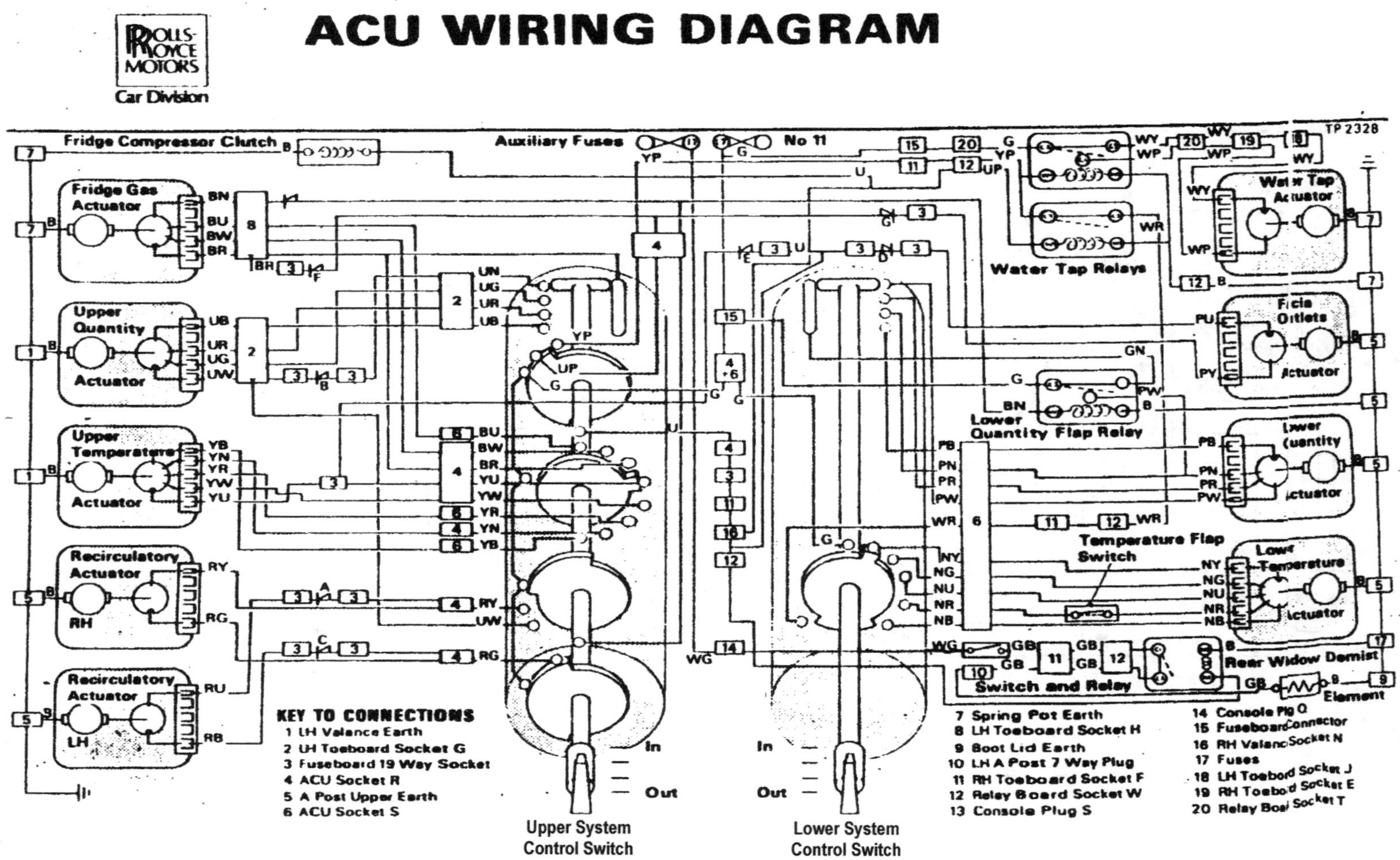

Fig. 8-4-6 – ACU Electrical Diagram
(Illustration supplied courtesy of Bentley Motors, Inc. Upper and Lower System Control Switch labels corrected by author.)

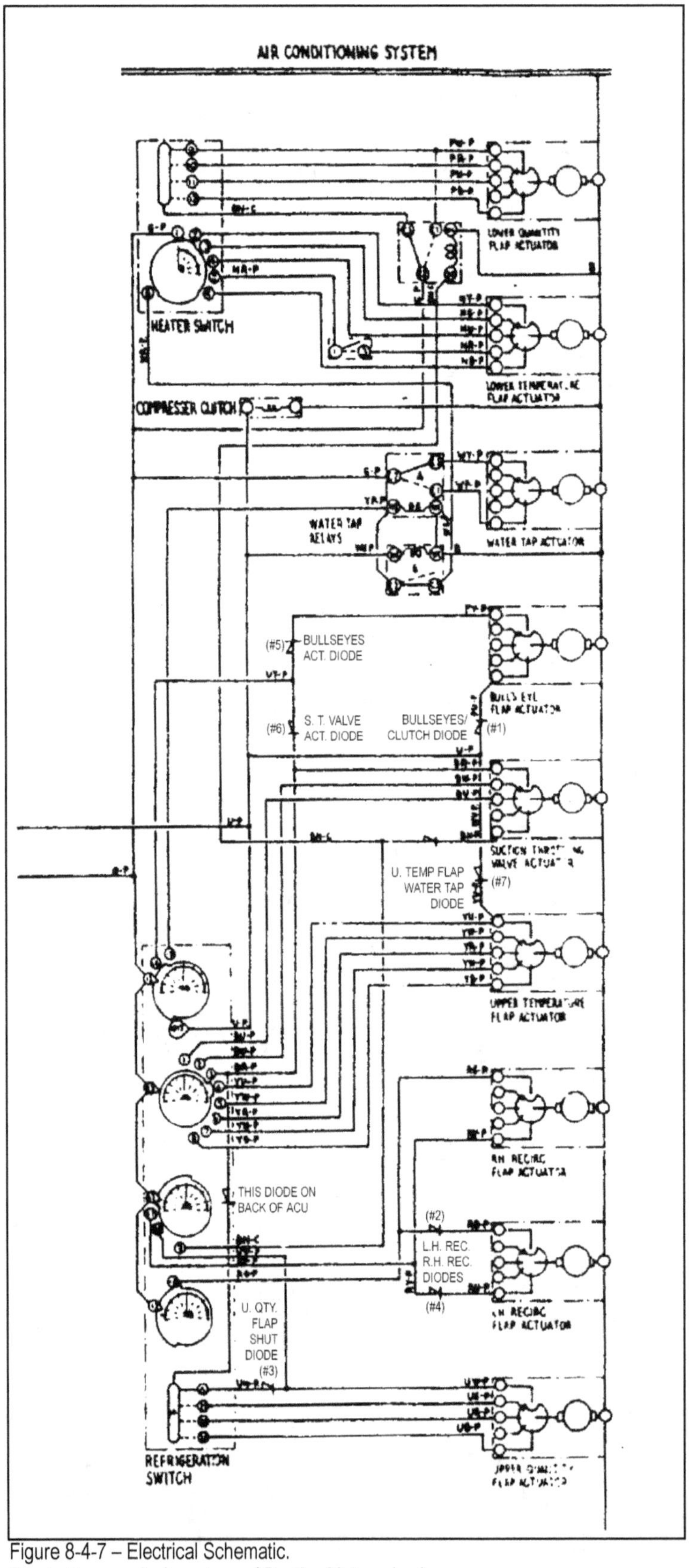

Figure 8-4-7 – Electrical Schematic.
(Illustration supplied courtesy of Bentley Motors, Inc.)

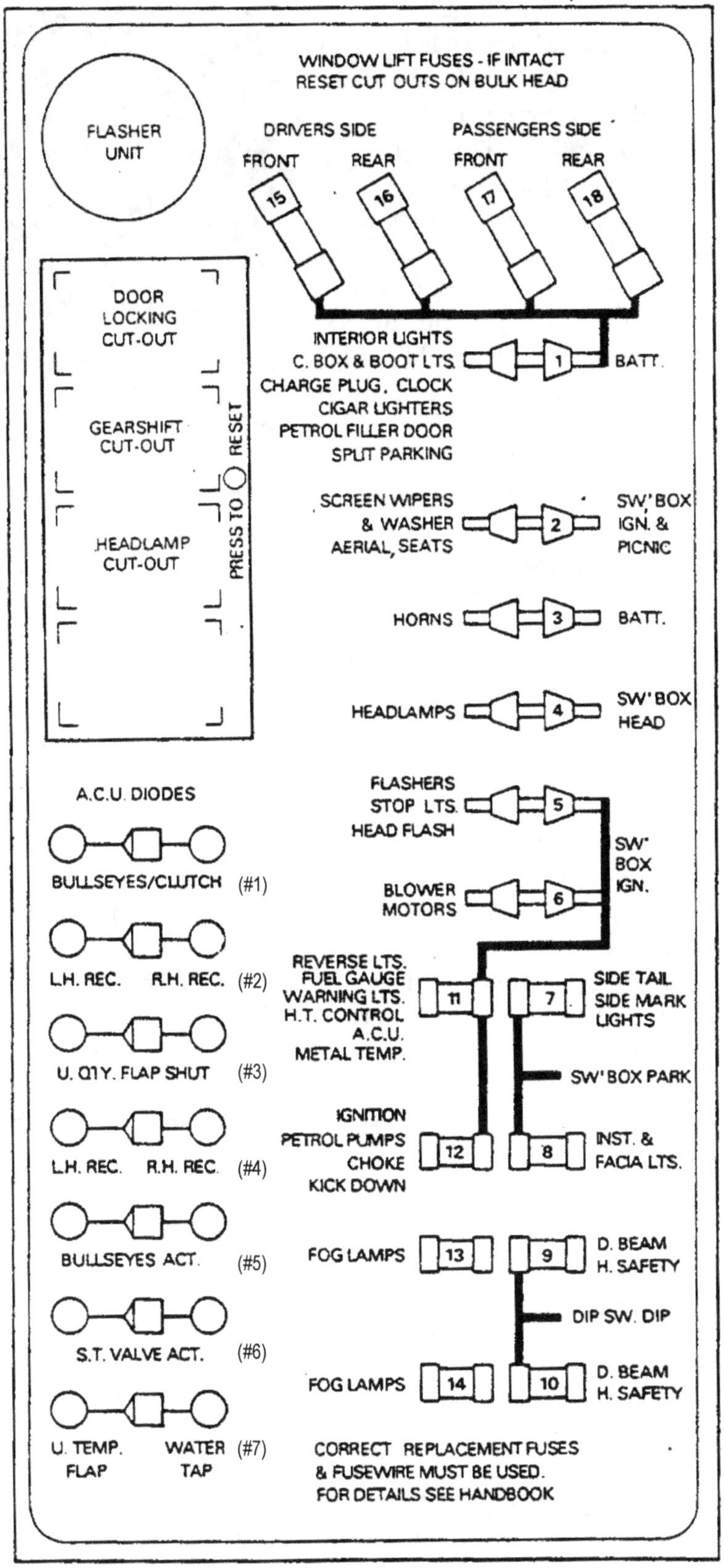

Figure 8-4-8 – Shadow I Fuse Board.
(Illustration supplied courtesy of Bentley Motors, Inc.)

ACU and associated actuators is shown in fig. 8-4-6. The schematic is shown in fig. 8-4-7. A series of seven 1N4002 diodes are used to prevent backflow of current. Backflow can cause some interesting malfunctions that are very difficult to trace. As mentioned above, switching to maximum air conditioning will override the upper quantity setting and force that door closed to prevent windshield fogging. Without diodes in this circuit, backflow would also power the circular flap actuators continuously moving from fresh to refrigerated and back again! Fortunately, diodes fail to an open circuit and are much easier to diagnose.

All of the diodes in the fuse board are arranged with the cathode on the left and the anode on the right. Since diodes are unidirectional conductors, the orientation is important. The cathode side is marked with a white stripe on the diode. Diodes must be removed from the fuse board to test them. (See fig. 8-4-9.)

A 7BA socket will remove the retaining nuts. Using a multimeter with a diode test feature, the voltage should read 0.5-0.7V when forward biased (red wire to anode and black wire to cathode) and infinity when reverse biased. 1N4002 rectifier diodes are very common and can be purchased at Radio Shack. In a pinch, 1N4003 and 1N4004 diodes will also work, but not look quite the same in the fuse board. Note: there is one diode for the air handling unit that is not on the fuse board, it is located on the back of the ACU.

There are many possible failure modes for the air handling system. The following three examples cover the gamut. Understanding these will enable you to diagnose just about anything the system can throw at you.

<u>*Lower (or Upper) Temperature Inoperative*</u>

In this case, warm air came from the register on the floor, regardless of the lower temp setting. Changing the setting produced no movement of the lower

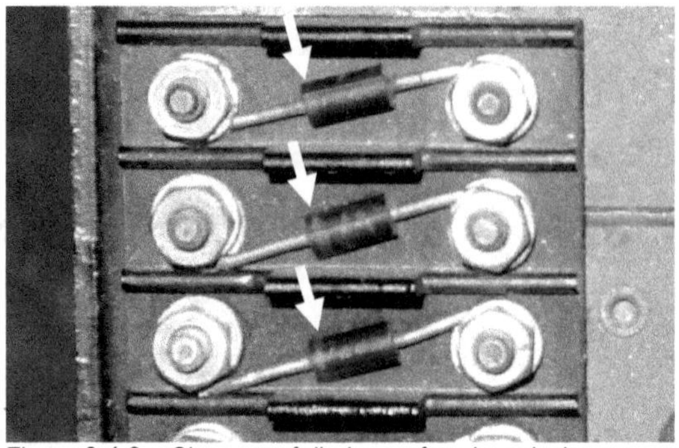

Figure 8-4-9 – Close-up of diodes on fuse board. Arrows point to white stripe which identifies cathode end.

temperature servo actuator, which is located on the left side of the center console. Probing the lower servo's electrical connector showed the voltage was not changing from pin to pin as the setting on the temperature knob was changed. Clearly, the problem in this case lies in the ACU.

To gain access to the ACU on center console equipped Shadow I's; remove the two triangular side panels with a tape covered screwdriver. They are held in place with the same reusable metal clips used to secure the door panels.

Remove the 8-track player knobs and remove the ash tray. Pull the seat adjuster switch assembly back just far enough to allow the center console wood to pitch backward to clear the 8-track player and then lift it out.

Unscrew the ACU from the front, disconnect it from the wire harness and then ease the unit through the side of the center console.

When the cylindrical dust covers are removed from the ACU, the cause of the problem in this case was obvious: the rotating disc that controls temperature had fractured and no longer held fast to the shaft. (See fig. 8-4-10 and 8-4-11.) New ACUs currently cost $1,271.

Fellow club member Tom Kindler came to the rescue of the car featured in this section with a salvaged unit for a much lower cost. When I received it, I found my unit to be in better cosmetic condition, so I took the parts I needed

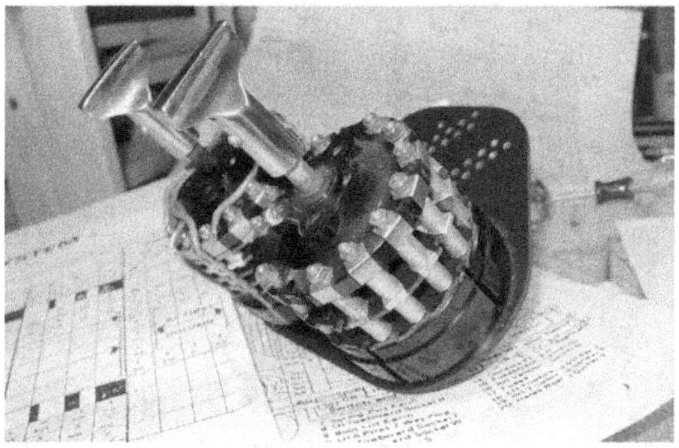

Figure 8-4-10 – ACU with dust covers removed.

Figure 8-4-11 – Fractured ACU Disc

from Tom's unit and reconditioned my original. This is not for the faint of heart; the retrofit involved the removal and installation of dozens of 7BA nuts, with excruciating attention paid to maintaining the proper location of all the contacts relative to their respective tracks.

<u>Left (or Right) Recirculation Inoperative</u>

There are two recirculation flaps located in each of the front foot wells against the bulkhead. They have a propensity to fall off their hinges. Diagnosis in this case was very basic: cycle the upper temperature knob between fresh air and air conditioning with the key on and observe operation of the flap. If it operates, yet you can move the flap by wiggling the actuating rod or it is stuck in an awkward position, you have found your culprit.

To address this, first remove the carpeted kick panel. It is nailed to a wooden strip on the leading edge of the door

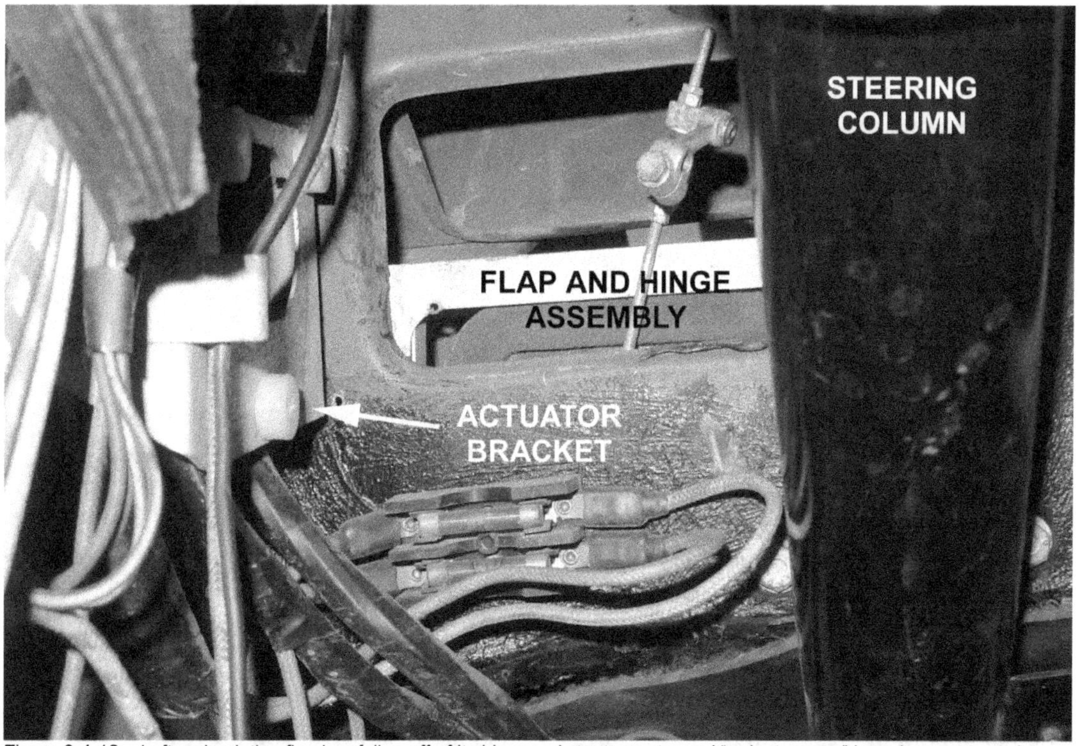

Figure 8-4-12 – Left recirculation flap has fallen off of its hinges. Actuator motor and "rodent screen" have been removed.

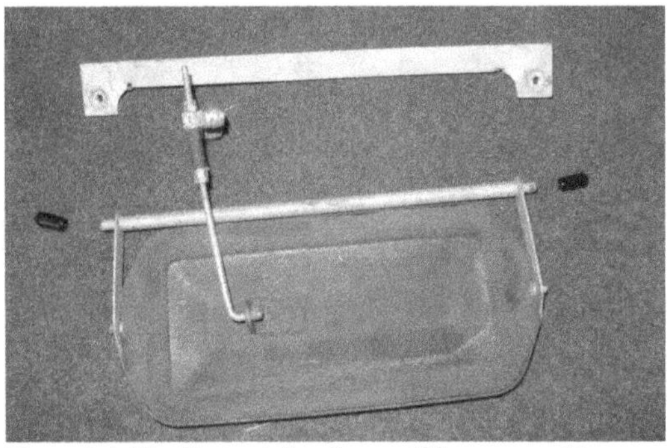

Figure 8-4-13 – Recirculation flap and hinge assembly.

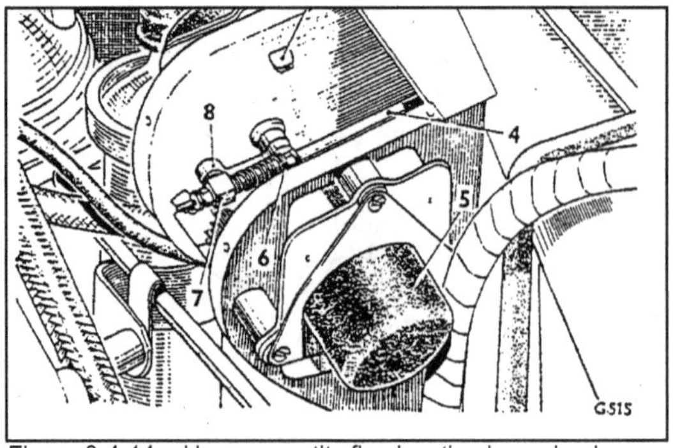

Figure 8-4-14 – Upper quantity flap location in engine bay. (Illustration supplied courtesy of Bentley Motors, Inc.)

aperture with a few brads. Disconnect the linkage at the shaft of the actuator by loosening the 2BA pinch bolt. Remove the actuator. Remove the front and rear lower bracket bolts. Loosen, but do not remove the front upper bracket bolt. Remove the four self-tapping screws that retain the "rodent screen" over the recirculation flap aperture. Remove the two bolts that thread through the bulkhead into the weld nuts on the recirculation flap hinge. (See fig. 8-4-12.)

On the engine compartment side, disconnect the blower motor wires and remove the three 2BA nuts that retain the blower fan motor. Remove the blower motor. Remove the three 2BA nuts that hold the blower fan housing to the bulkhead. Manipulate the flap and hinge assembly until it can be removed and pulled though the aperture into the cabin side of the bulkhead. (See Section 8-1 for further details on removing the blower motors.)

I have used #10 plastic screw covers to serve as new bushings to reattach the flap to its hinge, but fortunately the Crewe part is now available after a long hiatus. (See fig. 8-4-13.) Reassembly requires a second set of hands to hold the hinge against the bulkhead from the engine side while the bolts are threaded into it from the cabin side. Once everything is back together except for the actuating rod, turn the upper knob to maximum heat with the key on and allow the servo motor to rotate to the fresh air position. Reattach the actuating rod to the actuator without tightening it. Pull the flap shut against the bulkhead and ensure that the lever on the servo actuator is parallel to the actuating rod. Tighten the pinch bolt.

Upper Quantity Inoperative

Sometimes the servos will react slowly, grinding to a halt without reaching the desired position. This particularly affects servos with several harness connections such as the upper quantity servo. (Location shown in fig. 8-4-14.) The cause of this type of issue is usually electrical.

Due to inherent voltage drops in the system, especially across the diodes, one should expect 9-11V at the live pins on the actuator connector. If you find readings on this connector more in the neighborhood of 6V or less, the search is on for poor connections.

As figs. 8-4-6 and 8-4-15 illustrate, this actuator is fed through the left hand toe board connector. The engine side of this connector is located directly below the left hand blower motor housing mentioned previously. It is not well protected and corrodes over time, increasing resistance. All of the upper quantity wires go through connector G, which is the connector closest to the upper quantity actuator. Quick removal of this connector and a light cleaning with some brake cleaner will usually restore the voltage readings to a more respectable level.

When this occurred on my car, cleaning the toe board connector as described restored operation of the upper quantity flaps in the A/C mode, but not in the fresh air and heat modes. It turns out that if diode #6 on the fuse board, labeled as "S. T. VALVE ACT. DIODE" in fig. 8-4-7, fails to an open circuit you will be unable to select upper quantity in the heat and fresh air modes. This diode passes voltage to the upper quantity push-pull switch when fresh air is selected. Replacing this diode corrected the problem.

As far as I am concerned, the diodes labels are all a bit misleading. You can waste an extraordinary amount of time troubleshooting the A/C system, chasing shadows thinking that all of the relevant diodes have been tested. The first thing I do when faced with an actuator failure is remove and test *all* of the diodes regardless of their supposed function. 1N4002 diodes are cheap and readily available. Stock a few spares!

Figure 8-4-15 – Left hand toe board connector visible with blower motor and housing removed.

Chapter 9
Miscellaneous

Section 9-1
Tools and Supplies for a Road Trip

I always found myself clamoring to get all the necessary items for an overnight trip in my Silver Shadow at the last minute. You don't necessarily need to carry along all of these things, just make sure one of the other cars in your caravan does. Electronic copies of this list are available the Sherbourne Mews website.

Supplies

2 Spark Plugs
One Long Spark Plug Wire
Spare Ignition Coil
Duct Tape
Electrical Tape
Points (non Opus)
Points File (non Opus)
Engine Oil
Oil Funnel
Transmission Fluid
Transmission Fluid Funnel
RR363 Brake Fluid
Rags
Premixed Coolant
Coolant Funnel
Alternator Belt
Gloves (Mechanix® original style are nice.)

Niceties

Cell Phone
Phone Charger
Sunglasses
Sunblock
Folding Chairs
Cleaning Supplies
Camera(s)
Digital Camera Charger 12V and 120V, TV and Computer Cords, Memory Card
Wine Opener
Name Tag(s)
2-Way Radios
Umbrella, Hat
Beverages and Cooler
California Duster

Tools

3/8-in. Ratchet
3/8-in. Slider Bar
3/8-in. Ratchet Extensions
3/8-in. SAE Sockets
3/8-in. Universal Joint
Mini Ratchet Set (Griot's Garage no. 10247)
1/4-in. Ratchet Extension
1/4-in. SAE Sockets
BA Sockets
7/16-in. and 11/16-in. Wrenches (1/2-in. and 9/16-in. in Factory Toolkit)
BA Wrenches
2BA Nut Driver
Timing Light
Engine Analyzer (With Dwell Function)
IR Thermometer
Digital Multimeter with Diode Test Function
Wiring Diagram
Fire Extinguishers
13/16-in. Spark Plug Socket
Leatherman® Tool
Cordless Utility Light
Nylon Tie Down Straps for Four Wheels
Heavy Duty Jumper Cables
Magnetic Pickup Tool

Section 9-2
Tire Selection

Since the beginning of the Shadow era, tire nomenclature has changed along with tire design. For those of us that own early examples, a bit of mathematics is required to arrive at a modern tire recommendation in current nomenclature.

The Workshop Manual recommends "205 15" radial-ply tires for all cars prior to the chassis numbers listed in fig. 9-2-1 and "235/70 HR 15" radials for those listed and thereafter. The breakpoint corresponds to the introduction of flared wheel arches midway through the 1974 model run.

Model	Chassis No.
Silver Shadow	SRC 18269
Bentley T	SBH 18265
Long Wheelbase	LRH 19577
Corniche Convertible	DRH 18563
Corniche Saloon	CRH 18564

Figure 9-2-1 – Tire Selection Breakpoint (see text)

Flared Arch Cars

Purchasing 235/70HR15 radials is straightforward as the specification is still in use. Note that the speed rating is "H" which corresponds to 130 mph. Although this handily exceeds the car's top speed, it is important to go with this speed rating or higher because it provides an extra good measure of sidewall stiffness to cope with the relatively low inflation pressures specified, as well as a higher resistance to force balance issues that lead to high speed vibration.

Non-flared Arch Cars

This size is an outdated Euro-metric specification. If you were to convert this to current Euro-metric nomenclature, it would read 205/82R15, which is not readily available. The 205 indicates that the width, sidewall to sidewall, is 205mm. The aspect ratio is 82%, which means that the height of the tire from the tread to the rim is 82% x 205mm or 168mm. The diameter of the rim is 15 inches or 381 mm. Thus the overall tire diameter is 168mm x 2 + 381mm = 717mm.

The largest aspect ratio passenger car tire readily available today is 75. A 205/75R15 tire would calculate to an overall diameter of 689mm which is 4% smaller in diameter than the original tire. Although the width is ideal, the smaller tire would cause the speedometer to read 4% too fast as well as appearing to be "lost" in the wheel well.

The best compromise is a 215/75R15, which is an indistinguishable 2% smaller in diameter. The additional 5% in width looks "right" and does not interfere with any suspension components when steering wheel is turned to lock to lock. Rolls-Royce did not specify a speed rating, S (112 mph) and T (118 mph) rated tires are readily available in this size and I would recommend the higher of the two.

Load Index

As seen on the tire sidewall, the tire size is usually followed by the load index. The load index corresponds to the load carrying capacity of the tire. The Shadow is a heavy car relative to the size of the tires so it is important to pay attention to this number as well. For either series of cars, go with a load index of at least 100. The load index usually has the speed rating reiterated as a suffix. Thus, the load index should be 100T or better on non-arch Shadows.

Noise

The largest variation owners seem to experience between brands is noise. Talk to other Shadow owners and to find out what the current recommendation is. For quite a while it was a now discontinued line of Michelins.

Age

Tires degrade whether you use them or not because of their exposure to ozone and ultraviolet radiation. Consider changing them every six years. For recommendations in a nutshell, please refer to fig. 9-2-2.

Non-flared Arch Shadows
215/75R15
Load index 100T or better

Flared Arch Shadows
235/70HR15
Load index 100H or better

Figure 9-2-2 – Summary of tire recommendations.

Section 9-3
Battery Replacement

When it comes time to replace your battery, do so with a "Group 27" battery. Group 27 is a Battery Council International standard that governs the size of the unit and location of the terminals. In this case, the battery is 12-1/2 x 6-13/16 x 9 inches. Your battery purveyor may need to order a battery in this size, but he or she will instantly recognize what you are talking about when you specify the group size.

When specifying a battery this size, the battery will usually have a rating of 700 cold cranking amps (CCA) or so, which is perfectly adequate for all but the most extreme conditions. (If you plan to leave your Silver Shadow parked outside for the month of January at Detroit Metro Airport, I certainly do not want to hear about it!)

Purchasing the correct battery will ensure a proper fit within the battery box with a snug fit underneath the battery hold down bracket and the terminals in the expected locations. (See fig. 9-3-1.)

Battery hold down brackets tend to rust when they are exposed to battery acid. They often go missing entirely for this reason, or because the battery size is incorrect. This is not a good thing!

The bracket is fairly easy to source second hand and can be sand blasted and repainted. I would recommend powder coating it satin black. The coating is extremely durable and you can have it done for around $25. (See fig. 9-3-2.)

Figure 9-3-1 – Proper battery with cover removed. (Illustration supplied courtesy of Bentley Motors, Inc.)

Figure 9-3-2 – Powder coated battery hold down bracket.

Section 9-4
Build a Manometer

A liquid column manometer measures the amount of applied pressure, positive or negative, relative to atmospheric pressure. Manometers of this type use mercury or water depending on the sensitivity required. When pressure is applied to the manometer the level of the mercury or water changes in direct proportion to the pressure applied. Measurements relative to classic automobiles are usually in "inches of mercury (Hg)" or "inches of water."

You will most often use this instrument on your Silver Shadow to measure the vacuum in the carburetor float chambers. The tubing is sized to connect directly to the pressure tapping nut on the "A" side carburetor float chamber. Since float chamber vacuum is 3.25 inches of water or less, the manometer you are constructing will need to use water. (The equivalent pressure reading to 3.25 inches of water is 0.24 inches of mercury—too small to measure, not to mention the health risks of handling mercury!)

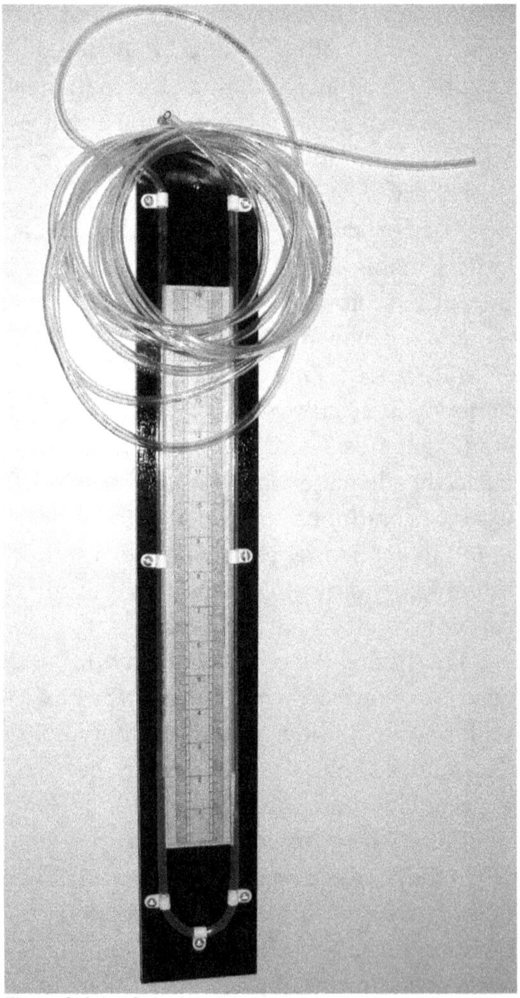
Figure 9-4-1 – Completed manometer.

Procure the items listed in fig. 9-4-2. Paint the poplar board as you see fit and glue the paper scale to the board. Using a 3/16-in. drill bit, drill seven holes in the approximate positions shown in fig. 9-4-1. Place the holes so that the screw mounted plastic clamps will support the tubing so that it runs in a "U" with the tubing parallel on each side of the paper scale.

Assemble the tubing to the board using the clamps, screws, nuts and washers. (Washers and nuts go on the back side of the board.) Arrange the tubing so that one end of the "U" is flush with the top of the board and the excess tubing exits the opposite side of the "U". This is the side you will connect to the engine.

Mount the manometer to a wall in your garage so that it is perfectly vertical and within reach of your Shadow's engine. Add distilled water to fill the bottom three inches or so of the "U".

Refer to fig. 9-4-3 when using your manometer.

Qty	Item
1	24 x 1/4 x 2-in. poplar board
1	20 ft. coil, 1/4-in. I.D. plastic tubing
7	#8-32 x ½-in. machine screws
7	#8-32 nuts
7	#8 washers.
7	1/4" plastic clamps (Found in electrical section—used to secure phone wire.)
1	Paper scale in inches (Download one for free at sherbournemews.com)
-	Distilled water to fill the bottom three inches or so (Distilled water evaporates without residue.)

Figure 9-4-2 – Manometer bill of materials.

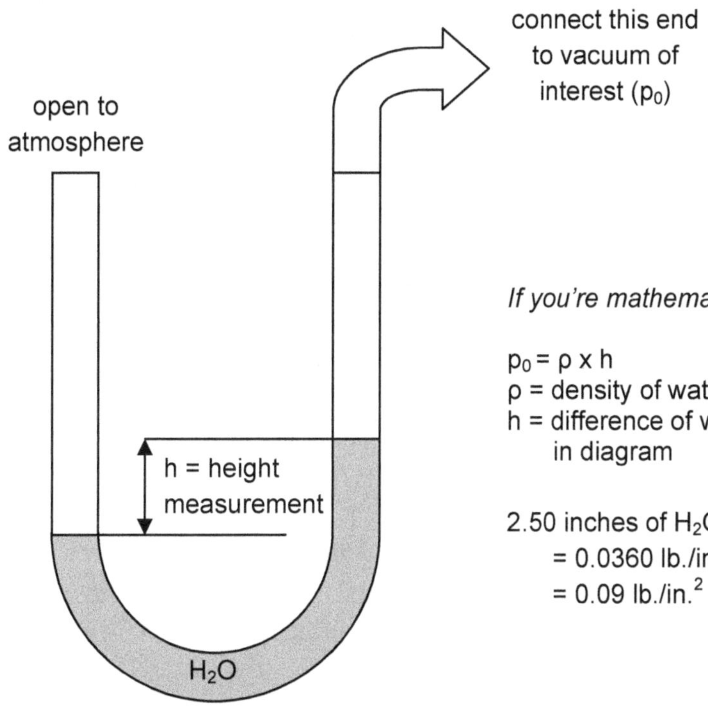

If you're mathematically curious:

$p_0 = \rho \times h$
ρ = density of water (0.0360 lb./in.^{3})
h = difference of water height as shown in diagram

2.50 inches of H_2O
= 0.0360 lb./in.^{3} x 2.50 in.
= 0.09 lb./in.^{2} or 0.09 psi

Figure 9-4-3 – How to use the manometer.

Section 9-5
Troubleshooting Electrical Circuits

<u>*Basic Electrical Terminology*</u>

Voltage or electrical potential is the difference in electrical charge between two points. It is expressed in Volts. The voltage difference between the two terminals on a properly charged modern car battery is about 12.8 Volts, or 12.8 V.

Current is the rate of flow of electric charge over time. It is measured in Amperes. When referring to smaller amounts of flow, *milliamperes* are used instead. One milliampere is one thousandth of an Ampere. When your rear window defrost circuit is switched on, it draws approximately 7 Amperes, or 7A.

Resistance is the measure of resistance to the flow of electric charge. It is measured in Ohms which is often abbreviated with the Greek letter omega (Ω). A kilo-Ohm or KΩ is 1,000 Ohms. Resistance elements or resistors are often placed in electrical circuits to regulate the flow of current to other devices. For example, a ballast resistor reduces the current to the ignition coil so that it does not overheat.

All components in an electrical circuit have resistance to a degree. A long run of heavy gauge wire may only be 0.1 Ω, whereas a low beam headlamp has a resistance of 3.5 Ω. The resistance of an open switch is infinite, allowing zero current to pass.

<u>*Ohm's Law*</u>

Voltage, current and resistance in a circuit behave very predictably and are described by this fundamental formula:

V = I x R

V = voltage
I = current
R = resistance

For example, if you measure the voltage across the terminals of your heated rear window to be 14 V and measure the current through it to be 7 A, then the resistance must be 2 Ω. (If Voltage equals current times resistance, then resistance equals voltage divided by current. 14 V divided by 7 A equals 2 Ω.)

The map light circuit is shown in simplified form in fig. 9-5-1. To measure the voltage you would set your multimeter to "V_{DC}" and measure by touching a probe to

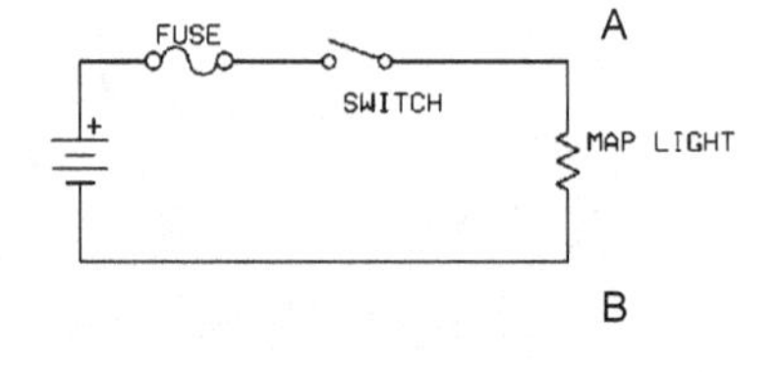

Figure 9-5-1 – Simplified map light circuit.

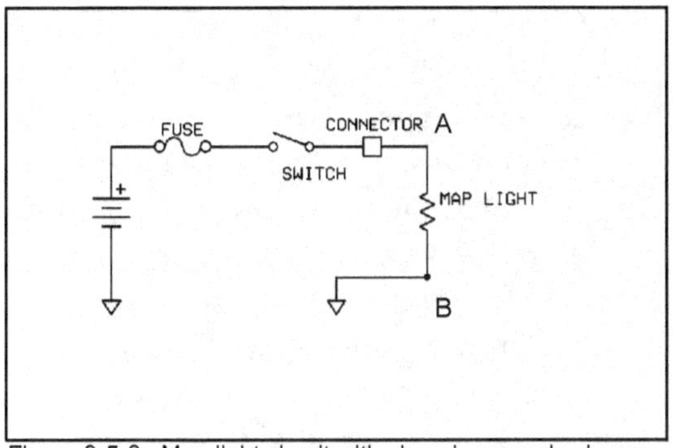

Figure 9-5-2– Map light circuit with chassis grounds shown.

Figure 9-5-3 – Bakelite fuse holder with fuse wire in place.

each of the metal terminals (at points A and B) while allowing everything to remain connected.

To measure the current, you would set your multimeter for "Amps," disconnect terminal A and touch one probe to the disconnected wire and one to the connection on the map light.

Negative Chassis Ground

You've often heard the terms "Negative Ground" and "Positive Ground" or as the British like to say "Negative Earth" and "Positive Earth." Since your Silver Shadow is made almost entirely of conductive metal, it would be pretty silly to run two wires to every component in the car. The Silver Shadow is negative ground, which means that the negative terminal of the battery is solidly connected to every large metal bit on the car, including the engine. The only post-WWII Rolls-Royce or Bentley models with positive ground are the Silver Dawn, Mk VI and R-type.

The fuse board underneath the steering wheel is connected to the positive terminal of the battery. Each major circuit is connected to a fuse on the fuse board and then run to an on/off switching device and then to the device itself which has one side connected to the switch and the other side simply bolted to the chassis to complete the circuit. (Fig. 9-5-2 shows the circuit diagram enhanced to reflect this.)

Fuses

On chassis numbers 27000 and higher, the fuse board consists of conventional Buss® type fuses, a flasher unit and two circuit breakers. The fuses are easily replaced (always with the same rating) with fuses purchased at any automotive parts retailer.

For chassis numbers earlier than 27000, there is a row of fuses that are identical to those on my 1939 Wraith. Each one consists of a reusable Bakelite holder which holds a strand of fuse wire between two copper clips.

To replace the wire in one of these fuses, remove the fuse from the fuse board by pulling on it. Gently remove the copper clips. Remove the remnants of the old fuse wire. Cut a suitable length of fuse wire from the spool, which is also located in the fuse board. Wrap one end of the wire around one of the side protrusions on the holder. Then run the wire across the holder to the protrusion on the other side. Replace the copper clips ensuring they come into firm contact with the fuse wire. Install the fuse. (See fig. 9-5-3.)

Although there are exceptions on other models, including my pre-War Wraith, under no circumstances should you run multiple lengths of fuse wire across the fuse holder or substitute heavier gage wire on the Silver Shadow. If the fuse keeps blowing, it means that there is something wrong.

Connections

The electrical system is connected by a series of wire bundles. These bundles are made in chunks that are easy to install when the car is being manufactured. They are connected to each other at various locations throughout the car. The main connection locations are each kick panel underneath the dash ahead of the front doors, and the toe board connectors passing through the bulkhead underneath each blower motor. These connections can become corroded or dirty leading to inoperative components.

In addition, there are junction blocks in each door and discrete connections for all of the exterior lamps. These also become corroded and dirty.

Measuring Voltage

Set the multimeter to "V_{DC}". DC is an abbreviation for "direct current." The other option is alternating current, which is what you find in your house. If you inadvertently set your meter to "V_{AC}" and you measure DC, the meter will read zero. (See fig. 9-5-4.)

To measure the voltage drop across a component, simply touch each probe firmly to each of the metal connections while the component is switched on. The red lead should be on the positive side of the component and the black lead should go to the ground side. If there is only one wire going to the component, touch the red lead to that and the black lead to a nearby piece of metal.

If you happen to reverse the leads on a digital multimeter, the reading will merely be a negative number. If you do this on an old fashioned style multimeter, the needle will go out of range to the left of zero and you will not be able to get a reading.

Measuring Current

Measuring current is usually not required for troubleshooting. To measure current, the multimeter often requires you to move the red probe to a different connection point on the meter. Consult the instructions that came with your particular meter.

You may also need to set the meter to read current in the "mA" range or perhaps the "A" range depending on what it is you are about to measure. (Many digital multimeters are autoranging. Consult the instructions that came with your meter.) Headlights, heated backlights and other ancillaries draw many amps so the larger

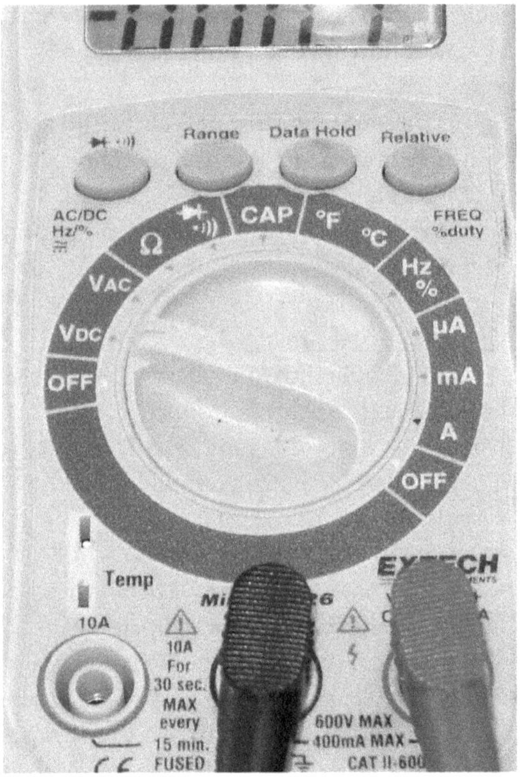

Figure 9-5-4 – Typical autoranging digital multimeter.

setting should be used. If you are measuring drain from the clock or something of this nature, use the smaller "mA" setting.

To measure, disconnect the device to be measured from power. Connect the red probe to the powering circuit and the black probe to the device to be measured. The reading will indicate the current draw through the device being measured.

Do not attempt to measure large current draws such as those found in the starter motor circuit. These circuits carry over 100 Amperes and can kill you, not to mention fry your meter or at least, blow its fuse. The same goes for the Alternator charging circuit.

Measuring Resistance

Set the multimeter to Ohms or "Ω". When in this mode, your multimeter applies a tiny, precise voltage across the probes and then measures the resultant current. It applies Ohm's law and displays the resistance value. If you hold the probes apart, the reading is either infinity, a flashing screen, or a seriously high number in the megohm or MΩ range. If you touch the probes together, the reading should be zero.

Testing Diodes

Diodes are like an electrical version of a check valve. They allow current to flow in one direction only. Most of the diodes are found on the fuse board adjacent to the steering column in the passenger cabin. Diodes on the Shadow are usually used to prevent unintended behavior in the air conditioning system.

Diodes cannot be tested in place. To test a diode, remove the two 7BA nuts and washers affixing it to the fuse board. Select the diode test feature on your multimeter (just to the left of "CAP" in fig. 9-5-4) and hold the black probe to the lead closest to the white stripe on the body of the diode. This is called the *cathode.* Attach the red probe to the opposite lead. This lead is called the *anode.* The meter should read approximately 0.5 V.

Reverse the probes so that the red lead is now attached to the cathode and the black lead is attached to the anode. The meter should read in a manner consistent with an open circuit. The reading when you have the probes attached to the diode in this

Figure 9-5-5 – Close-up of diodes on fuse board. Arrows point to white stripe which identifies cathode end.

direction is no different than holding the two probes apart in the air.

If the diode fails either of these two tests, replace it. They are inexpensive, plentiful and available at Radio Shack. On Shadow I applications, all of the diodes used are standard part number 1N4002.

Diagnostic Technique

When diagnosing an inoperative component, proceed in two paths at once: from most likely to least likely and from easily checked to difficult to check.

If the component is inoperative, suspect the component first. If it's a map light, for example, replace the bulb and see if the new one works. If the component is something for which there is no immediate replacement, test it with a 12V power supply on the bench or run battery power to it directly.

If the component is good, then it must not be receiving power. First, use your

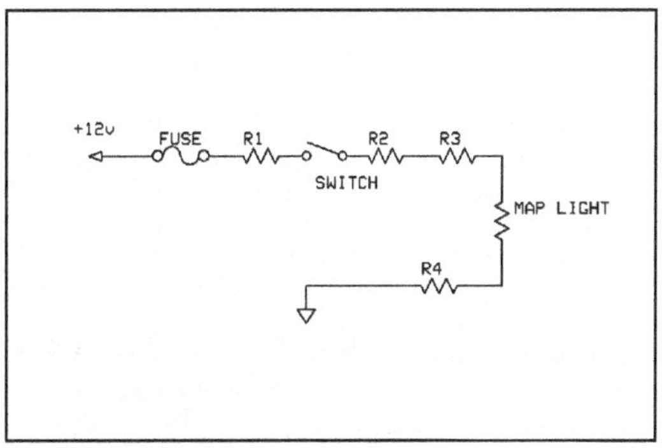

Figure 9-5-6 – Map light circuit diagram.

multimeter to verify that the part is grounded properly. In this case, the resistance between the component (in this case a map light) and ground is represented by "R4" in fig. 9-5-6. To check if it is grounded properly, measure the resistance between the negative ground lead connection and a solid metal, unpainted portion of the chassis. The multimeter (in the resistance mode) should read less than 0.5 Ω. If there is no ground lead, check the resistance between the metallic portion of the component in question and a solid ground on the body.

If the ground is good, check for power. With the component in place, measure the voltage where the positive lead attaches to the component. Attach the red probe to the positive lead on the component and the black probe to a clean chassis ground. If the voltage is less than 12 V, move up stream, as it were, toward the positive terminal of the battery. You are searching for bad connections that add resistance (and voltage drops) in the circuit. These are represented by R1, R2 and R3 which, for argument's sake, are the connector behind the fuse board, the connection at the map light switch and the kick panel connector, respectively.

In general, switches and wires do not fail very often. Usually it is the individual connectors adjacent to the component or the wire harness connector blocks found in each door, the front kick panels on both sides and the toe board underneath the blower motors that become corroded and resistive.

Because resistive connections create a voltage drop in direct proportion to the current going through them (Ohm's Law), it is important to check voltage at various points in the affected circuit with everything in place so that current can flow through the circuit. If the light bulb is removed from the circuit, the voltage at the positive terminal might read close to 12V instead of the 3V it would read with an operational current draw.

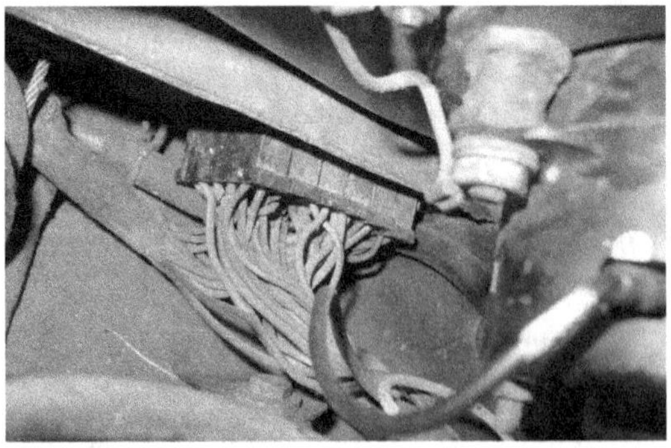

Figure 9-5-7 – Toe board connector beneath blower motor.

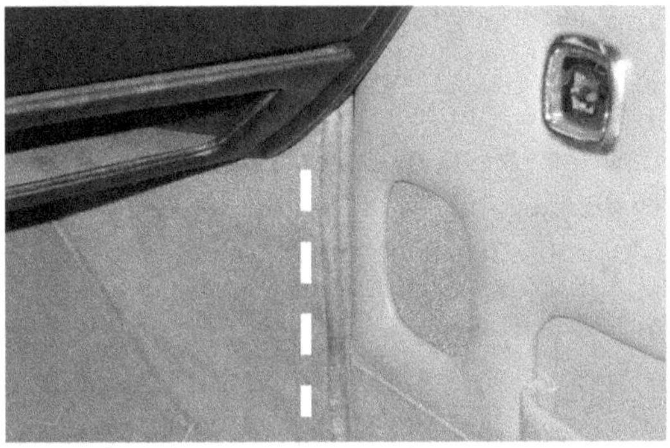

Figure 9-5-8 – Location of kick panel tack strip.

If you are serious about diagnosing electrical issues on your Silver Shadow, you will need a copy of the wiring diagram for your particular year. There is a degree of variation from the first cars through the last with additional permutations for the coachbuilt variants. It is not as scary as it sounds: a full car electrical schematic easily fits on one 11x17-in. page.

Connector Block Locations

All connections from the passenger cabin to the engine compartment pass through toe board connectors located beneath the blower motors on the left and right hand sides of the bulkhead. (See fig 9-5-7.) These connections include the upper temperature and quantity actuators, suction throttling valve, heater tap, lighting, warning lamps, gear change actuator, blower motors and resistors, horns, and A/C compressor, to name a few! These connectors are manufactured by Lucas and are known as LUCAR connectors.

All of the connections that run from the dash to the doors, such as window, door lock, cigar lighter power, and speakers run through LUCAR connectors behind the left and right kick panels. (See fig. 9-5-8.)

Most exterior lamps have single or dual connections. (See fig. 9-5-9.) Replacements for these connections are available from your Crewe Parts dealer. Headlamp and front parking lamp connections are neatly nestled in a block immediately behind each headlamp pair.

Figure 9-5-9 – Single connector.

Chapter 10
Resource Guide

Section 10-1
Sources for Parts and Services

203 Custom Car Works, 309 N. Prospect, Ypsilanti, MI 48198, 734-483-5772
Restoration paint and body work

Acme Car Service, 1015 Grove Way, Hayward, CA 94541, 510-582-6800
Service, advice and salvage Rolls-Royce parts
www.acmecar.com

Bentley Zionsville, 360 S. First St., Zionsville, IN 46077, 317-873-2360
Service, advice and Crewe parts
www.albersrollsbentley.com

British Starters, 225 Horton Highway, Mineola, NY 11501, 516-978-6161
Denso starters
www.britishstarters.com

British Tools and Fasteners, 7696 Route 31, Lyons, NY 14489, 315-946-9400
BA fasteners and taps
www.britishfasteners.com

Burlen Fuel Systems Ltd., Spitfire House, Castle Road, Salisbury, Wiltshire SP1 3SA, United Kingdom, +44-1722-412-500
SU carburetor and fuel pump parts
www.burlen.co.uk

Constantines Wood Center, 1040 E. Oakland Park Blvd., Ft. Lauderdale, FL 33334, 800-443-9667
Wood stains, veneers
www.constantines.com

D&D Custom Bright Works, Inc., 1311 B South St., Piqua, OH 45356, 937-773-5127
Chrome plating

Dun-Right / Q.C. Coatings, 50710 Rizzo Dr., Shelby Township, MI 48315
Powder coating
www.qccoatings.com

Eastwood Co., 263 Shoemaker Road, Pottstown, PA 19464, 800-343-9353
Buffing equipment, Simichrome polish
www.eastwood.com

Griot's Garage,
Car care products for the anal retentive
www.griotsgarage.com

Gunson, Kineton Road, Southam, Warwickshire CV47 0DR, United Kingdom, +44-1926-815-000
Colortune
www.gunson.co.uk

Healey Brothers, Ltd.
Replacement body panels
Merged with IntroCar Ltd. Sept. 18, 2006. (See IntroCar Ltd. listing below.)

Hilborn Motor Car Interiors, 7025 Reseda Blvd., Reseda, CA 91335, 818-345-2113
Reproduction interiors

IntroCar Ltd., No. 1 Manorgate Rd., Kingston upon Thames, Surrey KT2 7AW, United Kingdom, +44-2085-462-027
Crewe parts
www.introcar.co.uk

Kingsborne Wirewerks, 1786 La Costa Meadows Dr., Suite 102/103, San Marcos, CA 92078, 800-643-0375
Ignition components
www.kingsborne.com

Leatherique Restoration Products Ltd., 106 Englewood Road, Aiken, SC 29803, 877-395-3366
Leather refinishing supplies
www.leatherique.com

McMaster-Carr, P.O. Box 94930, Cleveland, OH 44101, 330-342-6100
Fabrication supplies and tools
www.mcmaster.com

Newport Auto Center, d.b.a. Bentley Newport Beach, 445 E. Pacific Coast Hwy., Newport Beach, CA 92660, 949-673-0900
Crewe parts
www.newportautocenter.com

RROC Club Stores, 191 Hempt Rd., Mechanicsburg, PA 17050, 717-697-3671
Rolls-Royce technical and niche publications, R-R tool rental, BA sockets and wrenches
www.rroc.org

Sherbourne Mews, 3231 Sherbourne Rd., Detroit, MI 48221-1814, 248-327-4330
Niche automotive publications, one of a kind tools
www.sherbournemews.com

Tony Handler, 2028 Cotner Avenue, Los Angeles, CA 90025, 310-473-7773
Salvage Rolls-Royce parts
www.tonyhandler.com

Section 10-2
Clubs

Rolls-Royce Owners' Club Inc., 191 Hempt Rd., Mechanicsburg, PA 17050, 717-697-4671, 800-879-7762
www.rroc.org

Modern Car Society, an affiliate of the RROC for enthusiasts of Shadow era and newer cars, address same as above.
www.rroc-mcs.org

Rolls-Royce Enthusiasts' Club, The Hunt House, Paulerspury, Northants, NN12 7NA, United Kingdom, +44-1327-811-788
www.rrec.co.uk

Rolls-Royce Owners' Club of Australia
www.rroc.org.au

New Zealand Rolls-Royce and Bentley Club Inc., 46 Windemere Rd., Cristchurch, New Zealand, +64-3-352-1037
www.nzrrbc.co.nz

De Mascotte (Netherlands)
www.demascotte.nl

Rolls-Royce & Bentley Club of Japan
www.rrbc.jp

Section 10-3
Model Year Specific Tune-up Data

Model Year	Spark Plug Gap	Contact Gap and Dwell	Ignition Timing	Idle Speed	Weakener Vacuum	CO Setting
1967 SRX1001+	0.025 in.	0.015 in.	3° BTDC @ 500 RPM	500 RPM	-	-
1968-69 SRX6000+	0.025 in.	0.015 in.	0° BTDC @ 500 RPM	550-600 RPM	2.75-3.25" @ 2,000 RPM	4-4.5% @ 550-600 RPM
1970-71 SRX9000+	0.025 in.	0.015 in. 26°-28°	5° BTDC @ 800 RPM	600 RPM	2.75-3.25" @ 2,000 RPM	5.5-6% @ 600 RPM
1972 SRA	0.025 in.	0.015 in. 26°-28°	5° BTDC @ 800 RPM	600 RPM	2.75-3.25" @ 2,000 RPM	5.5-6% @ 600 RPM
1973 SRB	0.025 in.	0.015 in. 26°-28°	5° BTDC @ 800 RPM	600 RPM	2.0" @ 2,500 RPM w/ EGR off	5-5.5% @ 600 RPM
1974 SRC	0.025 in.	0.015 in. 26°-28°	15° BTDC @ 1,200 RPM	600 RPM	2.2" @ 2,500 RPM w/ EGR off	4.5-5% @ 600 RPM
1975 SRD	0.030 in.	0.011-0.017 in. Electronic	15° BTDC @ 1,200 RPM	600 RPM	2.0" @ 2,500 RPM w/ EGR off	3-5% @ 600 RPM
1976 SRE	0.030 in.	0.011-0.017 in. Electronic	15° BTDC @ 1,200 RPM	600 RPM	2.0" @ 2,500 RPM w/ EGR off	3-5% @ 600 RPM
1977-80 SRF-SRL	0.035 in.	0.012-0.017 in. Electronic	15° BTDC @ 1,200 RPM	650 RPM	2.0" @ 2,500 RPM w/ EGR off	1-4% @ 650 RPM
1980 Calif. SRL - C	0.035 in.	0.012-0.017 in. Electronic	15° BTDC @ 1,450 RPM	650 RPM	N/A	0.5-0.7% @ 650 RPM

Section 10-4
Good Reading

Bennett, Martin. *Rolls-Royce and Bentley: The Crewe Years.* Haynes Publications, June 2000.

Bobbitt, Malcolm. *Rolls-Royce Silver Shadow Bentley T-Series Camargue & Corniche.* Veloce Publishing, September 2004.

James, Brendan, Anthony Bird, and Ian Hallows. *The Rolls Royce Motor Car: and the Bentley Since 1931.* Batsford, March 2003.

Robson, Graham. *The Rolls-Royce and Bentley: Shadow, Corniche, Camargue 1965-1985: A Collector's Guide.* Motorbooks International, February 2000.

Robson, Graham. *Rolls-Royce Silver Shadow: The Complete Story.* Crowood Press, June 1998.

Wade, G.R. *SU Carburetters Tuning Tips & Techniques.* Brooklands Books, Ltd., September 1994.

West, Cal. *From the Shadow's Corner.* RROC Club Stores.

Wood, Jonathan. *Rolls-Royce & Bentley: Spirit of Excellence.* Haynes Publications April 2002.

Workshop Manual: Rolls-Royce Silver Shadow (including Long Wheelbase Saloon), Rolls-Royce Corniche, Bentley T Series and Bentley Corniche. Rolls-Royce Motor Cars Limited, 1988.

Workshop Manual: Rolls-Royce Silver Shadow II. Rolls-Royce Motor Cars Limited, 1988.

Illustrated Parts Catalogue 1965-1980. (Compact Disc) Rolls-Royce Motor Cars Limited, 1999.

Index

About the Author

Born in Bloomsburg, Pennsylvania, Jon Jeffrey Waples is the older of two children. He spent the majority of his childhood in East Greenwich, Rhode Island where his parents still reside today.

In 1988, he received a Bachelor of Science Degree with High Distinction in Electrical Engineering from Worcester Polytechnic Institute. After three years at Westinghouse Electric Corporation in Baltimore, he pursued a degree of Master of Business Administration from the University of North Carolina at Chapel Hill, which he received in 1993.

For the majority of his career, Mr. Waples has served in a variety of powertrain related research and development positions in the automotive industry including a seven year stint with Ford Motor Company.

Mr. Waples is a frequent contributor of technical articles for the Rolls-Royce Owners' Club periodicals *Modern Lady* and *International Lady.* He also sits on the Boards of Directors for both the Modern Car Society and the Motor Region of the Rolls-Royce Owners' Club.

He lives in Detroit with his partner Garry and Labrador Retrievers Dave and Katie. His stable of interesting cars is currently limited to two: a 1973 Rolls-Royce Silver Shadow and a 1939 Rolls-Royce Wraith Limousine by Park Ward.

Join now and receive the Modern Lady 6 times a year!

The Modern Car Society

An Affiliate of

The Rolls-Royce Owners' Club, Inc.

The Modern Car Society is a non-profit organization formed to address the interests of owners and enthusiasts of the Rolls-Royce and Bentley motor cars produced after 1965. Ownership of a Rolls-Royce or Bentley is NOT required as a condition of membership. Current membership in the RROC is required.

- Membership Application -

Name:____

First　　　　　　　　　　　Last

Street Address:______________________________________

____

City　　　　　　　State/Province　　　　　　ZIP Code

Spouse:____

First　　　　　　　　　　　Last

Home Phone:_________________ **Office Phone:**_________________ **Ext:**________

Fax:_____________________ **email:**___________________________

MOTOR CAR INFORMATION:

Car#1/Year: ______ **Make:** ______________ **Model:**____________ **Body Style:** ________

Chassis No: ______________________ **Engine Number:** ____________________

Car#2/Year ______ **Make:** ______________ **Model:**____________ **Body Style:** ________

Chassis No: ______________________ **Engine Number:** ____________________

PLEASE MAKE YOUR CHECK PAYABLE TO:

The Modern Car Society

Annual Dues: $30.00 (add $10.00 for postage if you reside outside of North America)

$15.00 of dues are for Modern Lady subscription. $1.00 additional for spouse/partner to be included with your MCS membership.

Mail Completed Membership Application to:

Rolls-Royce Owners' Club, 191 Hempt Road Mechanicsburg, PA 17055

1-800-TRY-RROC

www.ingramcontent.com/pod-product-compliance
Lightning Source LLC
LaVergne TN
LVHW080049210926
848481LV00011B/3309
9780615141039